Information Security and Cryptography
Texts and Monographs

T0223410

Giampaolo Bella

Formal Correctness
of Security Protocols

With 62 Figures and 4 Tables

 Springer

Author

Giampaolo Bella
Università di Catania
Dipartimento di Matematica
e Informatica
Viale Andrea Doria 6
95125 Catania, Italy

giamp@dmi.unict.it

Series Editor

Ueli Maurer
Inst. für Theoretische Informatik
ETH Zürich, 8092 Zürich
Switzerland

ACM Computing Classification: C.2.2, D.2.4, D.3.1, F.3.1, F.4.1, I.2.3

ISSN 1619-7100
ISBN 978-3-642-08782-0   e-ISBN 978-3-540-68136-6

Springer is a part of Springer Science+Business Media
springer.com

© Springer-Verlag Berlin Heidelberg 2007
Softcover reprint of the hardcover 1st edition 2007

Cover design: KünkelLopka, Heidelberg

*To the loving memory of my father, Carmelo,
whose death helped me broaden my views of life.*

*To my mother, Agata,
whose loss of eyesight helped me bring those views
to a focus.*

Foreword

This book describes a key technique, the Inductive Method, for proving the correctness of security protocols. It is clearly written, starting with the basic concepts of cryptography and leading to advanced matters such as smartcards and non-repudiation. The book is also comprehensive and timely, with some of the cited papers still in their journal's publishing queues.

Security protocols are short message exchanges designed to protect sensitive information from being stolen or altered. Mobile phones, Internet shopping sites and subscription television boxes all rely on them. Even with good cryptography, security protocols are subject to many types of attack. Perhaps a hacker can combine pieces of old messages to create what appears to be a valid response to a challenge. The danger is greater if the participants in the transaction can be expected to cheat, as with many Internet purchases. Because the number of possible attacks is infinite, the only way we can be sure that a protocol is correct is by mathematical proof.

Researchers have been attempting to prove the correctness of various computer system components since the 1970s. Hardware can be verified to a great extent using logic-based tools such as binary decision diagrams (BDDs) and model checkers. Software seems much more resistant to verification; recent celebrated work using SAT solvers can only prove very simple properties. Security protocols are software; to be precise, they are concurrent algorithms based on cryptographic operations. Unusually, these algorithms include a threat model: one process is assumed to be an enemy with wide powers to capture and combine other people's messages.

Given these complicating factors, it is unsurprising that protocol verification became a long-standing open problem in computer security. As late as 1995, the problem appeared to be intractable. Today, however, security protocols can be verified automatically using a number of freely available tools. Though many side issues remain, the core problem must be regarded as solved. Many people can take credit for this remarkable success. In particular, Oxford's Security Research Group pioneered the use of model checkers to analyse protocols [142]. Their work has been widely copied.

However, model checkers do not prove properties: they search a finite space for possible flaws. Being automatic, they are excellent for debugging, but the failure to find a bug does not mean that none exist. The Inductive

Method described in the present volume takes some ideas from the Oxford group, such as their message primitives, and applies them in the context of proof. Industrial-grade protocols such as SSL/TLS, Kerberos and even the huge SET protocol suite [37] can then be tackled. Although these proofs are far from automatic, the effort needed to undertake them is considerably less than that required to design the protocol in the first place. Merely to derive an abstract protocol from the standards documents can take weeks, a task that automatic tools cannot escape.

Giampaolo Bella is amply qualified to write this book. He has been involved with the Inductive Method almost from its inception. He has worked to extend its scope, for example to smartcards, and to refine the types of properties it can express. These efforts also illustrate why the Inductive Method is still valuable in the era of off-the-shelf protocol verifiers. It can help explain how these verifiers work: some of them are based on formal models similar to those described in this book. Besides, off-the-shelf tools inherently have a limited scope: they are designed to solve problems of a particular sort. Bella has shown that the Inductive Method can easily be extended to new security environments, which can then be studied formally. Such work enables the development of a new generation of automatic tools, and so the field progresses.

Cambridge, February 2006   *Lawrence C. Paulson*

Preface

The era of computer networks has nowadays surpassed the era of computers. It is widely realised that computer networks ought to be secure for innumerable reasons, ranging from prevention of frauds to accountability of actions. For the sake of security, the network participants must take predefined steps called security protocols. This book treats proofs of correctness of realistic security protocols. A formal though intuitive setting underlies all proofs, which the reader can inspect to be convinced of their validity. A proof is nothing but evidence that each protocol step preserves a desired property.

This book is the result of approximately ten years of my research, started back in late 1996. I was beginning my Ph.D. at the Cambridge University Computer Laboratory from some of my supervisor's newborn ideas on analysing security protocols using simple induction. Larry Paulson called his inductive techniques the "Inductive Method." As inventor of the theorem prover "Isabelle," he was probably most interested in the underlying proof aspects, whereas I soon turned my interests to the actual security issues. With his support, I have devoted these years to making the Inductive Method mature for real-world protocols.

My Ph.D. thesis [23] firmly underlies these pages. I have added four new chapters to it and rewritten the rest, so this manuscript is entirely new. Writing in English was a big endeavour for an Italian mostly living in Italy, but my international friends reassure me that the outcome is readable — after all, they are friends indeed! I am confident that this book is a valid help to understand the entangled niceties of security protocols and their verification, for both teaching and research purposes.

In terms of verification, the reader will learn how to inductively deal with general protocol features such as timestamps, message reception, agents' knowledge, smartcards, non-repudiation, certified e-mail, and other advanced goals. In terms of actual protocols, several designs will be taken apart to the last message brick, including BAN Kerberos, Kerberos IV, Kerberos V, a smartcard protocol by Shoup and Rubin, a non-repudiation protocol by Zhou and Gollmann, and a certified e-mail protocol by Abadi et al. Transversal competence concerns the theorem prover Isabelle, which offers the necessary mechanical support, and the principles of prudence underlying robust protocols. I advance and demonstrate "goal availability," a principle directly aimed at guiding protocol analyses and ultimately at designing correct protocols.

Acknowledgements

I am extremely grateful to Larry Paulson for the perfect mix of knowledge, experience, and patience with which he supervised my Ph.D. research. My gratitude goes to my Ph.D. examiners, Mike Gordon and Peter Ryan, for their constructive criticism. Special thanks are due to Margaret Levitt, administrative secretary of the Cambridge University Computer Laboratory, for her bureaucratic assistance, and to Lewis Tiffany, former librarian of the Laboratory, for making the library a friendly environment.

I would also like to thank many colleagues. Colin Boyd offered constructive criticism to my goal availability principle; Dieter Gollmann clarified the relation between integrity and authenticity; Peter Honeyman unveiled a few technicalities of the Shoup-Rubin protocol; Gavin Lowe facilitated my understanding of authentication; Markus Kuhn illustrated many weaknesses of smartcards; and David Richerby proofread my thesis.

My gratitude also goes to all professors of the University of Catania, who welcomed me back after I got my Cambridge degree. I especially recall Domenico Cantone, for quickly acknowledging my security competences; Giovanni Gallo, for supervising my undergraduate final dissertation; Alfredo Ferro, for initially encouraging me to go to Cambridge; Elvinia Riccobene for transforming my initial impetus towards research into expertise; and Salvatore Riccobene, for solving many equipment problems.

My friends have played a major role throughout these years. Marco Aurisicchio, Valeria Baiamonte and Giovanni Farinella offered me a familiar atmosphere in the summer of 2005 in Cambridge, when Pietro Liò was an entertaining office mate; Stefano Bistarelli encouraged me to publish this book; Francesco Cassaniti has been a painstaking supporter; Cetty Di Maria and Daniela Pettinato have always believed in me; Dario Greco has been a great hardware consultant; Sandro Politi has constantly been a brother; Angelo Rovito has been an assiduous self-confidence reminder; much of my English proficiency is due to Alain Wolf's patience; Ilenia Tinnirello and her group at the University of Palermo warmly hosted me for a month. Erina Cocuzza was the person whom I was happiest to tell that this book would be published.

Sono profondamente grato a Zio Tanino, Zia Nerina e in particolare a mio fratello Giuseppe per aver riempito la mia assenza da casa con perizia e vitalità. Senza di loro non avrei potuto accettare di vivere all'estero.

Questo libro è dedicato al mio papà, Carmelo, la cui morte mi ha aiutato ad ampliare la mia visione della vita.

Questo libro è dedicato alla mia mamma, Agata, la cui perdita della vista mi ha aiutato a mettere a fuoco quella mia visione.

Catania, February 2006   *Giampaolo Bella*

Contents

List of Figures

List of Tables

1. Introduction

Communications across modern computer networks should be *secure*, an adjective that embodies multiple properties. For example, one may wonder whether a message just received was altered during its transfer. If not, then the message is said to enjoy *integrity*. Even if a message that is received quotes someone as its creator, he might be a fake one. If not, then the message conveys *authentication* of its creator. Another important property is whether the message that is received was intercepted and understood by others besides its creator and intended receiver. If not, then the message enjoys *confidentiality*. *Peer* generically refers to an endpoint of a remote communication. The peers belong to a set of *agents*.

Devising a complete list of properties to assure secure communications in the context of modern computer networks is matter of current research. Each property implicitly assumes the existence of a malicious agent, the *Spy*, whose aim exactly is to violate the communications profitably. The Spy can overhear messages during transfers, create fake messages and introduce them in the traffic. While history tells us that various forms of security have been important ever since ancient times, significant frauds have been orchestrated in recent years with the help of computers containing inexpensive hardware and software.

Cryptography may help. Used extensively also during the World Wars [97], it is the art of coding and decoding information by means of a *cryptographic key*. A cleartext message is transformed into a ciphertext one using the key (through an operation that is called *encryption*). In the best case, the cleartext can be retrieved from the ciphertext (through an operation that is called *decryption*) if and only if the key is available. In consequence, the cleartext is safe from the Spy as long as she does not know the key. On the contrary, the intended receiver of the message is assumed to know the key. When the cryptographic key used for encryption is the same as that to be used for decryption, cryptography is said to be *symmetric* or *shared-key* (DES [125], IDEA [105]); otherwise, it is *asymmetric* or *public-key* (RSA [138], LUC [149]). A *digital signature* uses techniques of asymmetric cryptography to confirm the author of a digital message. A *message authentication code, MAC* in brief, for a message is another message computed using techniques of symmetric cryptography from the original message and a key that the peers share.

This brief outline of cryptographic terminology signifies that the underlying mathematical foundations are not of specific interest to this book. Additional readings are easy to suggest ([97, 119, 146, 150]).

Steganography [98] may also be used for secure communications. It is the art of hiding a message inside a larger, intelligible one so that the Spy cannot discern the presence of the hidden message after seeing the larger one. For example, the low-order pixel bits of a digital image may be changed to the bits of a message to be sent confidentially while the image does not suffer perceptible variations.

A more recent technique aiming for confidentiality and authentication is called *chaffing and winnowing* [137]. It may be considered a form of steganography, but it makes use of MACs. Sender and receiver must initially agree on a secret key by using a key-exchange protocol such as Diffie-Hellmann [73] or Oakley [129]. The sender authenticates his message by computing the correct MAC for it and sending the pair formed by the message and its MAC off to the receiver. The sender also sends *chaff*, namely a large number of other pairs, each made by an intelligible random message and a wrong MAC. Only the intended receiver of the message can discern which pair brings the correct MAC, as he knows the secret key used to compute it. So, he alone can *winnow* the received messages, namely discard the chaff and select the original message.

The vast majority of *security protocols* for computer networks are based on the first technique, cryptography; hence, they typically are *cryptographic protocols*. These are sequences of steps that pairs of remote peers must take to subsequently establish a secure communication session between themselves. Each step requires the transmission of a *message*, possibly encrypted, between the peers. Messages include peer names, cryptographic keys, random numbers, timestamps, ciphertexts and concatenations of those components. A security protocol attempts to achieve certain *goals* at the time of its completion, namely a set of security properties.

Experience shows that security protocols often are flawed in the sense that they fail to enforce their claimed goals. A security protocol precisely is a concurrent program that can be executed by a large population of agents including the Spy. Not only is the Spy entitled to participate in the protocol as any other agent, but she can also act illegally, interleaving a number of protocol sessions. By doing so, she can exploit on a session the messages obtained from others. Moreover, the vulnerabilities of current transport protocols let her overhear the messages exchanged by other agents. Using security measures such as cryptography to enforce the protocol goals in this setting is not easy. This claim is supported by the large number of flaws that have been reported. To only mention a few flaws, some affect well-known protocols [13, 106], others can be classified [151]. Another group affects less publicly known banking protocols, whose weaknesses have been exploited by dishonest employees [12]. Other flaws are due to specific implementations of the cryptographic primitives [143].

These few citations confirm that establishing whether a protocol lives up its promises may be very difficult. This process was only carried out by informal reasoning until the late 1980s. If a protocol claimed to achieve a goal, some researchers studied the protocol in detail and decided whether this was true. At present, informal reasoning still retains its importance:

- it is crucial for grasping the semantics of protocol designs beyond their bare representation as message sequences;
- it may find minor weaknesses or simple flaws in a protocol more quickly than formal reasoning;
- it is fundamental for understanding certain flaws thoroughly;
- it is easier to follow by an inexperienced audience;
- it ultimately helps for developing formal reasoning and understanding formal guarantees.

While informal reasoning was failing to capture serious protocol flaws, the early 1990s saw increasing awareness that formal reasoning can be conducted profitably on abstract protocol models [58, 99, 117]. It can effectively prove a protocol correct in a model or otherwise detect realistic flaws. As we shall see (Chapter 2), some methods of conducting formal reasoning lack expressiveness or automation, others are just too complicated to use on realistic protocols or to suscitate industrial interest.

This book is about the use of the *Inductive Method*, which is supported by the theorem prover *Isabelle*, to formally prove correctness of realistic protocol models. While the foundations of the Inductive Method are due to Paulson [133], our aim exactly is its development to make it capable for real-size protocols. In achieving this aim, we have considerably extended the method, deepened the formal reasoning about protocols and ultimately developed a general principle of prudent protocol analysis. Therefore, this book can be profitably read by at least anyone interested in any of the following:

- understanding the entangled technicalities hidden behind various types of security protocols;
- learning a method of conducting formal analysis of realistic security protocols;
- teaching (verification of) security protocols;
- practicing with the theorem prover Isabelle;
- practicing with a general principle to realistically conduct formal analysis of security protocols.

An excellent companion to the present manuscript is Boyd and Mathuria's recent book on security protocols [55], which eminently discusses a variety of protocols and their underlying philosophy using a precise though informal language. By contrast, our book uses a formal language to systematically disassemble the protocol features down to the smallest component and bring to light details that might otherwise remain hidden.

The organisation of this chapter is simple. First, the motivation to our work is discussed (§1.1), and our contribution to knowledge is sketched (§1.2). Then, the notation that will be used throughout the book is presented (§1.3), and the remaining chapters are outlined (§1.4).

1.1 Motivation

The foundations of the Inductive Method date back to 1996, and are presented in Chapter 3. This section refers to that initial development stage, when our research and the subject of this book initiated. Hence, the present tense here refers to late 1996. Our motivation is threefold: the development of a young and promising method; the verification of real-size protocols that have never been formally explored in a realistic setting; and the investigation of general principles underlying correct protocols.

To gradually introduce the Inductive Method, here is its main underlying idea: simple mathematical induction suffices to model security protocols and reason about their goals. A key concept is the *trace*, a list of network events occurring while an unbounded population of agents is running a protocol. Traces are defined inductively and so is the set of all traces admissible under a specific protocol. This set represents the formal protocol model. Proofs can be carried out by induction on a generic trace of the model, with mechanical support offered by the theorem prover Isabelle. They establish trace properties representing goals of the underlying protocol.

1.1.1 Developing the Inductive Method

Further testing. A general theory of messages and an extendible formalisation of the Spy are already available in the method. An important feature is that no bound is stated on the size of the models. Crucially, the population of agents who could participate in the protocol is potentially infinite: the model agents originate from a bijection with the natural numbers. Also, each agent is allowed to interleave an arbitrary number of protocol sessions.

However, the method has only been applied to a few classical security protocols [131, 132]. To convince ourselves of the practicality of mathematical induction in this context, further case studies are necessary. The security community appears to lament that the size of the existing case studies is not realistic. Hence, we choose to turn our attention to largely deployed protocols and to intrinsically different protocols, such as non-repudiation ones.

Deeper understanding. Other informal criticism of the method derives from difficulties in accepting the concept of trace, considered a "low-level" view of the network traffic, or a structure that is "non-existent" in reality.

A trace can be viewed as a possible *history* of the network events occurring while the protocol is executed. This interpretation may help us understand the key concepts of the method, although it should be verified over additional case studies.

All proofs follow the natural inductive style adopted by humans: verifying that the various protocol steps preserve a certain property. However, proofs may seem "cryptic" and, consequently, their results may be accepted with reluctance. This is often due to streamlined proof scripts, which feature highly automatic proof methods implementing several proof steps. We intend to favour human inspection of proofs by often preserving the linear application of the proof steps.

Additional elements. Many protocols use timestamps to assure freshness of important components such as session keys. The current datatype of messages does not feature timestamps. A related open issue is how to express *freshness* in terms of timestamps.

The reception of protocol messages is not modelled. However, understanding a formal guarantee very often involves some informal reasoning about reception. Thus, it seems desirable to treat this event formally, although it is not obvious whether the existing analyses would be simple to update accordingly.

Another important issue is how to account for e-commerce protocols, which often involve smartcards. How could the cards be modelled, taking into account the risks of cloning? How could their functionalities and interactions with the agents be represented? Along the same lines, non-repudiation protocols are expected to require a non-standard threat model in which no agents trust each other. Its modelling price is not trivial to anticipate.

1.1.2 Verifying the Protocol Goals

Existing guarantees. The Inductive Method already features a general method for proving the goal of confidentiality, but the early literature [131, 132] fails to mention the concept of *viewpoint*. The formal guarantees about the protocols are expressed in terms of theorems established with Isabelle's support. They are useful to the protocol peers only when the peers can verify whether the theorem assumptions hold. For example, if a proof of session key confidentiality is available on assumptions that the protocol initiator can verify, then the protocol achieves confidentiality from the initiator's viewpoint. Unless the proof can be conducted also on assumptions verifiable by the responder, the protocol does not necessarily guarantee confidentiality to the responder. As an extreme concern, we may wonder whether a guarantee featuring assumptions that are impossible to verify would be of any practical importance.

While the treatment of confidentiality is, as mentioned, satisfactory, that of authentication is not. The latter is an important and complicated goal that may hold in a hierarchy of forms, as confirmed by Lowe using another formal method [109]. However, the current formalisation of authentication using the Inductive Method fails to express the knowledge of the very message components that authenticate the agents (a satisfactory explanation of the various authentication forms must be deferred until later, §4.6). Investigating how many and which forms our method captures at present is challenging.

Novel guarantees. The well-known goals of integrity, authenticity, key distribution and strong forms of authentication need to be treated formally with the Inductive Method. To illustrate, we observe that even when the initiator is informed that session key confidentiality holds, it is not consequently obvious that the responder shares the same session key or that he means to share it with the initiator. It is not clear at this early stage what and how substantial the extensions necessary to formalise these goals might be. Some of the existing guarantees might have to be reinterpreted.

1.1.3 Investigating the Protocol Principles

One of the ultimate goals of analysing security protocols formally is to derive the general principles that make them secure. Ideally, we would like to have simple rules, adherence to which would be easy to check and would directly guarantee the security goals.

The best-known set of principles of prudent protocol design is due to Abadi and Needham [7]. The main principle is *explicitness*, which prescribes each message to say exactly what it means without ambiguity. Other principles include avoiding unnecessary encryption and synchronising the network clocks when timestamps are used. Unfortunately, these principles are far from ideal, as they are neither sufficient nor always necessary to assure security.

It is interesting to investigate whether the findings obtained using the Inductive Method would support these principles or clarify how relevant they are to the goals of a protocol. Moreover, the deeper study of the goals that we advocated above might unveil new principles not only to design the protocols but also to analyse them realistically.

1.2 Contribution

Our contribution is multifaceted. It is simplest to present it in relation to the motivation of the research.

1.2.1 Inductive Method

The Inductive Method (Chapter 3) turns out to be easily extendible. Timestamps are modelled using a discrete formalisation of time based on the position of each event in a trace. Each history of the network is equipped with a global clock corresponding to the length of the corresponding trace, and so each trace has a global clock yielding the current time of the trace. All agents refer to it, so the model hides problems of clock synchronisation. A message component is considered fresh in a trace if the time interval between the creation of the component and the current time of the trace is less than or equal to the lifetime allowed for the component. Session keys are considered valid if and only if they are used within their lifetime.

These extensions have allowed us to mechanise the proofs of correctness of three protocols that make use of timestamps: the BAN Kerberos protocol (Chapter 6), the larger and deployed Kerberos IV (Chapter 7), and finally the more recent Kerberos V (Chapter 9). Although they share the structure of a few messages, each protocol hides peculiar subtleties. Their proof scripts are fairly intuitive because of the limited use of automatic proof methods, which is another of our goals.

New events can be modelled. In particular, introducing message reception makes the specifications more readable and the proofs easier to follow, increasing overall intuitiveness. The existing scripts can be updated pragmatically, with minor effort. Message reception is not forced to occur. This models a network that is entirely controlled by an active Spy who can intercept certain messages and prevent their delivery. Moreover, the reception event allows the formalisation of any agent's knowledge, rather than just the Spy's, in terms of message deducibility.

New elements, such as extra trusted servers or smartcards, are modelled as new types of the language. For smartcards, the interaction with their owners is formalised by additional events (Chapter 10). The agents' knowledge, in particular the Spy's, must be reviewed for two reasons: (i) all long-term secrets are now stored only in the cards; (ii) certain protocols that are based on smartcards assume that the Spy cannot listen to communication between a card and its owner, while other protocols do not make this assumption. We verify the entire Shoup-Rubin protocol (Chapter 11) using a faithful model obtained both from the informal specification of the protocol and the description of its implementation. The protocol involves new long-term secrets, which can be easily introduced in the definition of agents' knowledge.

Another success is the treatment of *accountability protocols*, which aim at giving peers evidence of each other's participation. They require a fundamental change to the threat model because the Spy no longer is an opponent between two peers who trust each other but, rather, can hide behind any of the peers (Chapter 12). We shall see that this change is not difficult to implement through the complete analyses of a non-repudiation protocol by Zhou and Gollmann and a certified e-mail protocol by Abadi et al. (Chapter 13). Some of these protocols assume the existence of a communication channel secured by a standard protocol such as TLS/SSL [72]. We have found simple formalisations for secure transmission over such channels using specific forms of the available events.

To summarise, our research equips the Inductive Method with all the necessary features to tackle industrial protocols. The subsequent verification of the SET protocol [36, 37, 38] confirms this. In general, the method has become so expressive that the bare statements of the theorems convey most guarantees without considerable informal argument to make on top of them. Therefore, the exposition accompanying each theorem reduces to the very minimum.

1.2.2 Protocol Goals

We find that the argument about any protocol goal can (and must, see the next section) be interpreted from the viewpoint of each peer, although this may not be trivial regardless of the formal method in use (Chapter 2). While this practice provides the human analyser with a better understanding of each protocol step, it also produces formal guarantees that the peers can apply in practice, as we shall see. During this interpretation process, we realise that one assumption of a theorem that has been proved for the shared-key Needham-Schroeder protocol is in fact entirely superfluous. Our study of the protocol goals supports the claim that the goals of authenticity and integrity are equivalent (§4.3), while the corresponding guarantees can be derived from reinterpreting some of the existing theorems (Chapter 4).

Paulson's method for proving confidentiality is still effective after the modelling of timestamps. However, several specific lemmas are necessary with Kerberos IV because of its hierarchical distribution of session keys. We have unveiled an important weakness in the protocol management of timestamps and lifetimes: it lets the Spy exploit certain session keys in realistic circumstances within their lifetime. Moreover, the agent to whom the session keys have been legally granted is no longer present on the network, and so will not register any irregularity.

Two different definitions of agents' knowledge are developed and variously compared. One may appear to be simpler as it exclusively relies on message creation through trace inspection. The other is based on full message deducibility from the traffic that is sent or received (Chapter 8). The latter requires an explicit formalisation of message reception, and so allows us to formally study the goal of key distribution. We argue that this goal is equivalent to a strong form of authentication (§4.7), as we demonstrate on both BAN Kerberos and Kerberos IV. As for Kerberos V, we shall see that its goals are somewhat equivalent to those of Kerberos IV, although its design calls for alternative proof methods.

Verifying the goals of the protocols that are based on smartcards only requires minor modifications to the existing proof methods. A set of simplification rules must be proved to deal with the new events and the new definition of agents' knowledge. Also the smartcards may require guarantees that the protocol goals are met. Two of the messages of the Shoup-Rubin protocol lack crucial explicitness, so that none of the peers knows which session key is associated with each other. The confidentiality argument is significantly weakened in the realistic setting in which the Spy can exploit other agents' smartcards. The proofs suggest a simple fix to the protocol, yielding stronger guarantees against veracious threats.

The goals of accountability protocols can be also analysed by induction. We have developed simple proof methods for the main goals of validity of evidence and of fairness. The former confirms that certain messages truly count as evidence of an agent's participation in the protocol. Fairness is an additional goal, requiring the appropriate evidence to be either available to both peers or to none, so that no one is advantaged. Various forms of both goals can be proved by simply showing that certain events always precede specific others on the protocol traces. This treatment is demonstrated on both the non-repudiation protocol and the certified e-mail protocol.

Admittedly, our proofs were difficult to develop. The analyses of Kerberos IV and Shoup-Rubin, for example, saw certain proofs take up to four man-weeks each to be developed, while the corresponding script was up to 50 Isabelle commands long. Polishing the original scripts often shortens them up to approximately one fifth of their original length, thanks to a moderate use of Isabelle's automatic proof methods, in which subsidiary lemmas can be installed. As mentioned, this must be done with care, for it may affect the resulting intelligibility. However, proof scripts are doomed to change over time as Isabelle evolves, while proof methods will rarely change. Hence, this book concentrates on the general methods rather than on the actual Isabelle scripts, some of which are demonstrated in the appendices.

1.2.3 Protocol Principles

Our research favours the development of protocol principles, namely those meta-rules that contribute to guaranteeing security. However, no principle is found to be sufficient in general.

To be precise, the importance of explicitness is confirmed. The messages that are not explicit about their meaning force the peers to heuristic interpretations that turn out to be extremely risky. This was known to affect classical protocols such as the public-key Needham-Schroeder; but we find that it also affects protocols that are apparently stronger, such as Shoup-Rubin, a smart-card one. We stress that studying adherence to the explicitness principle always requires an assessment of the underlying threat model. For example, if a communication is assumed to be preauthenticated, then quoting the peer names may be unnecessary.

Explicitness is also interesting from the proof perspective. We find that, if a message lacks explicitness, then carrying out any proofs about it requires quantifying existentially the exact components that are not sufficiently explicit. The prover needs either to bind them to the assumptions or to conjecture that they exist. It could be argued that mere expertise in theorem proving can discover lack of explicitness and therefore make up for competency of prudent protocol design.

The verification of Kerberos IV confirms the principle that extra encryption does not necessarily strengthen confidentiality. Despite the double encryption of the responder's session key, the key is vulnerable to the attack mentioned above, under a realistic threat model. Additional support for the principle derives from the analysis of Kerberos V, which attains similar confidentiality goals and suffers the same attack although it disposes with double encryption.

Our attention to the agents' viewpoints in carrying out formal protocol analyses leads us to the development of a principle of prudent protocol analysis. This seems to be the first time that a principle is spelled out to guide the analysis of protocols rather than directly their design. Secure protocol design of course remains the ultimate aim. But analysis and design are equally important because the former is meant to influence the latter. Only a realistic analysis will contribute to a truly more robust design.

Our principle of prudent protocol analysis is called *goal availability* (Chapter 5). It holds for a protocol, one of its goals and one of its peers if there exist guarantees from the peer's viewpoint that the goal is met. The guarantees must rely on assumptions that the peer is able to verify in practice. However, for each peer we can identify a set of assumptions that are necessary although the peer can never verify them: they form the peer's *minimal trust*. Goal availability tolerates the minimal trust.

Adherence to goal availability may be sufficient to prevent certain attacks or weaknesses either directly (as with Kerberos IV) or indirectly (as with Shoup-Rubin). The attack on Kerberos IV is in fact due to a violation of goal availability where the goal is session key confidentiality for the protocol responder. The weaknesses of Shoup-Rubin are due to the lack of explicitness discovered through the verification of adherence to goal availability. In this light, our principle seems easier to verify than the explicitness principle, whose definition is less constructive.

1.3 Notation

The notational conventions that are used throughout this book are summarised here. Although they are rather standard, digesting them appropriately is very useful to understanding this book thoroughly.

1.3.1 Presenting the Protocols

Getting to grasps with the syntax of messages is fundamental. Fat braces "{|" and "|}" [133, §2.1] are used to distinguish protocol messages from sets, and also to indicate the encryption operation. They are omitted in the case of messages whose outermost constructor is concatenation, and in the case of ciphertexts whose body is a single-component message. For example:

- a ciphertext made by encrypting with key K a two-component message consisting of m concatenated with n is indicated as $\{\!|m, n|\!\}_K$;
- a two-component message consisting of m concatenated with n is indicated as m, n;
- a single-component message m encrypted with key K is indicated as m_K.

The security protocols will be presented using what today is a rather standard notation. For each protocol step (which essentially sends a message), the step number, the sender and the intended recipient of the message, and the message itself are indicated.

$$
\begin{array}{llll}
1. & A \longrightarrow B & : & A, Na \\
2. & B \longrightarrow A & : & \{\!|Na, Kab|\!\}_{sK_B^-}
\end{array}
$$

Fig. 1.1. Example protocol

Figure 1.1 presents an example protocol that helps to demonstrate the notation. The protocol consists of two steps. In the first step, agent A sends agent B the two-component concatenated message formed with her own identity and Na. In the second step, agent B replies to A with a ciphertext obtained by signing with his key sK_B^- the concatenation of Na with another key Kab. It is unnecessary to detail the messages precisely at this stage. We in fact state nothing about the keys and only anticipate that Na and Nb are *nonces*. A nonce is a random "number that is used only once" [126]. In our example protocol, A invents the nonce Na: it has never been used before. Hence, its use tells A, upon reception of the second message, that B's reply is more recent than the instant Na was created. Nonces can also help establish various forms of authentication, as we shall see in this book.

A full understanding of the notation requires a reference to the *threat model* in which all protocols will be studied. The Spy can intercept all messages and prevent their delivery. She can also tamper with them by decomposing concatenated messages and opening up ciphertexts sealed with keys she knows. Then, she can use the learnt message components to form new messages at will by concatenation and encryption. This threat model, which we shall discuss more extensively (§3.9), is due to Dolev and Yao [75], and is a *de facto* standard for protocol analysis at present. In this threat model, a message that is sent is not necessarily ever received, and the receiver of a message is not necessarily its intended recipient. Therefore, a protocol step such as $A \longrightarrow B : A, Na$ signifies that A sends the two-component message to B but says nothing about B's reception of the message.

1.3.2 Naming the Theorems

This section presents a general naming paradigm that we adopt for all theorems throughout this book. The theorem names are identical to those in the proof scripts, which come with the Isabelle repository [33, 34] from the 2006 distribution. We shall see (§3.1) that it is useful to execute the relevant proof scripts interactively while the theorems are discussed throughout the book.

When we terminated a large protocol proof, which perhaps had taken up to weeks of concentration, one of the last things we usually worried about was the choice of the best, most expressive name for the theorem just proved. However, we now realise that a uniform naming system becomes especially important when putting together theorems about distinct protocols, as is the case with this book. It also helps in interpreting the theorems properly. Precisely, we have to face the following three issues.

1. *Theorem names should be correctly expressive.* When we read a theorem name, we would like to grasp its meaning, namely the goal it is trying to express, from its name. To achieve this, it was necessary to go back to all proof scripts and change many original theorem names. For example, the word "trust" was abused by many theorems establishing the originator of a message. Since that word is currently used mostly in relation with trustworthiness of agents, we found that it was inadequate. Authenticity of messages or authentication of agents seemed more appropriate terms. To only give an instance of this evolution of names, Theorem 6.5.2 is now addressed as **BK_Kab_authentic**, but its original name was **A_trusts_Kb2**. All theorem names are now coherent with the goal names that we set below (Chapter 4).

2. *Theorem names should be coherent between various protocols though not identical.* If we establish the same guarantee (say, of confidentiality) for more than one protocol, we would like the relevant theorems to have the same names so that they would favour comparative considerations. However, having two or more theorems named identically in a book may appear contradictory. To set about this issue, we decided to prefix *only in the book* each theorem name with the acronym in capital letters of the protocol name it refers to. In this vein, the prefix **BK_** of the mentioned Theorem 6.5.2 reminds us that it refers to the BAN Kerberos protocol. An analogous guarantee for Kerberos IV is Theorem 7.3.3, whose name **KIV_authK_authentic** correctly identifies the protocol. The different key name addresses the ambiguity only in this case. A more evident example derives from the session key confidentiality guarantees for Kerberos IV, such as Theorem 7.3.14 called **KIV_Confidentiality_B**, and for Kerberos V, such as Theorem 9.3.1 called **KV_Confidentiality_B**.

3. *Theorem names should be coherent between various versions of the same guarantee.* There are often various versions of the same theorem for various reasons. We maintain coherence between the various version names by adding a prefix and/or a suffix as follows.

- Once a theorem is proved, it is often possible to weaken or strengthen its assumptions and derive weaker or stronger facts. We indicate such a variant theorem by adding the _bis suffix to the original theorem name. For example, **SR_Outpts_A_Card_form_10_bis** is the name of Theorem 11.3.8. There may also be another variant, in which case the _ter suffix is added to the original theorem name.
- The fewer the assumptions made, the easier to grasp a theorem normally turns out to be. This is particularly so with assumptions of key confidentiality, which normally expand into several facts when relaxed. For the sake of presentation, it often becomes preferable to leave such assumptions unrelaxed. The theorem versions where all assumptions are relaxed into elementary facts are indicated by the _r suffix. For example, Theorem 6.5.10 is called **BK_A_authenticates_B_r**.
- Once a protocol is thoroughly analysed, we may find it relevant to update its design and repeat the analysis in a separate theory file. The theorem names for updated protocols receive a conventional prefix as described above. For example, Theorem 8.5.1 for our updated Otway-Rees protocol is called **ORB_analz_hard**, the **B** standing for Bella. Also the name of the new theory file is updated accordingly (§3.2). By contrast, the theorem names reflecting minor updates that can coexist within the original theory file are only continued with a **u**. Some of these theorems are mentioned but never presented in this book to limit redundancy.
- When we introduce the **Gets** event to formally model message reception (§8.2), the existing protocol models must be updated accordingly (in a separate theory file). The prefix of their theorem names is suffixed with a **g**. For example, Theorem 8.4.7 for the Kerberos IV model with message reception is called **KIVg_B_authenticates&keydist_to_A**. Also the name of the new theory file is updated accordingly (§3.2).

1.3.3 Wording the Symbols

Logical statements contain specific symbols with dedicated semantics. We replace most Isabelle symbols with the corresponding English phrases whenever their semantics is obvious and the replacement can improve readability.

- Logical conjunction ($\wedge$) with *and*.
- Logical disjunction ($\vee$) with *or*.
- Logical negation ($\neg$) with *not*.
- Logical equivalence ($\leftrightarrow$) with *if and only if*.
- Disequality ($\neq$) with *is not*.
- Meta-level implication ($[\![\ldots]\!] \Longrightarrow \ldots$) with *if ... then ...*.
- Existential quantification ($\exists$) with *for some*.
- Universal quantification ($\forall$) with *for any*.

By contrast, it is best to preserve other symbols exactly for the sake of readability. Isabelle's graphical interface offers perfect mathematical symbols [160].

- It is convenient to keep logical equality in symbols (=) because it can be used to specify variable expansions (abbreviations) among the theorem preconditions, while the variable can be compactly mentioned even repeatedly among the theorem conclusions. Isabelle supports well this form of equational reasoning, as this book confirms.
- Having conducted some experiments of wording set membership ($\in$), we concluded that it is quicker to grasp in symbols — with only one exception. It is preferable to state that *a list contains an element*, which is often needed below, rather than to use the symbol for set membership over the set of elements that the list contains.
- Other set operators, such as union ($\cup$) and inclusion ($\subseteq$), which are rarely used below, are kept in symbols.

Syntax is left completely unaltered when it is quoted for the sake of demonstration — always in figures — such as when the protocol models are presented.

1.4 Contents Outline

This section briefly outlines the chapters of this book.

Chapter 2 reviews some of the main formal methods for analysing security protocols. Such a large variety of methods has been advanced that the chapter cannot present all of them. The importance of interpreting the findings cautiously is emphasised. The difficulties in conducting the analyses do not seem to be related to those in interpreting the findings, as various examples confirm.

Chapter 3 outlines the Inductive Method as it was in late 1996, when our research initiated. The presentation of the method is gradual and informative, as it gives particular attention to the intuition behind each construct. The chapter begins with an introduction to the working environment for the method, namely the theorem prover Isabelle, and terminates with an example of a protocol model.

Chapter 4 formalises in the Inductive Method the most important guarantees for the protocol models. With every protocol that is analysed, those guarantees form the aim of our analysis, and hence the chapter introduces important terminology. Seven groups of guarantees are given, each expressing an important protocol goal, except for a group that helps to validate the protocol models. Proof methods and examples are provided.

Chapter 5 defines our general principle of prudent protocol analysis, goal availability. An abstract version is given first, in order to favour the reader's intuition, while a more detailed version only comes after additional discussion. Then, the related concept of minimal trust is put forward. The principle is demonstrated only on a simple protocol but additional examples are frequent in the subsequent chapters.

Chapter 6 extends the Inductive Method with a treatment of timestamping, which requires both a formalisation of time and a definition of timestamps. We find simple solutions for both issues. Then, the BAN Kerberos protocol can be formally analysed. Finally, a temporal modelling of the accidental loss of session keys is introduced. The protocol model is updated accordingly and all guarantees revisited correspondingly.

Chapter 7 presents the formal analysis of Kerberos IV. The treatment of timestamping and the temporal modelling of accidents is inherited from the previous chapter. The protocol achieves strong goals but it fails to conform to our principle of goal availability in the case of confidentiality of a group of session keys for the responder. This leads to realistic attacks, but a simple fix is introduced and verified.

Chapter 8 introduces the modelling of agents' knowledge in the Inductive Method using two definitions. They are demonstrated on the protocols presented above by adding treatments of the key distribution goal and of a stronger version of authentication, which were impossible before. The two definitions are variously compared and contrasted, offering an argument that can refute a claim previously made by the BAN logic.

Chapter 9 reports on the formal analysis of Kerberos V, the most recent version of the Kerberos protocol. The main difference with the previous version in terms of design is the removal of the use of double encryption. We verify that this difference does not significantly influence the protocol goals, which are analogous to those of the previous version. However, different proof methods become necessary.

Chapter 10 describes a realistic treatment of smartcards in the Inductive Method. Since some protocols explicitly assume that the communication means between the cards and their owners is secure, while others do not, our treatment develops around both options. The Spy is allowed to exploit an unspecified set of smartcards, some through simple theft, others through elaborate tampering ultimately leading to cloning.

Chapter 11 uses the extended method of the previous chapter to analyse the Shoup-Rubin protocol, which makes use of smartcards. The protocol is generally strong, but it is found to violate our goal availability principle for the goal of session key confidentiality. This reveals two important shortcomings of explicitness that affect three goals: confidentiality, authentication and key distribution. A simple fix is introduced and verified.

Chapter 12 extends the Inductive Method to deal with accountability protocols. Non-repudiation and certified e-mail delivery are recognised as forms of accountability, where the peers get evidence of each other's participation. Abstract formalisations of these novel goals are provided along with appropriate methods to prove them. The concept of second-level protocol, which relies on another security protocol, is advanced.

Chapter 13 uses the extensions introduced in the previous chapter to describe the analyses of two emblematic accountability protocols. They are both studied in terms of validity of the evidence provided to the peers and in terms of its fairness. The threat model appropriate for this group of protocols is used: an honest agent enjoys the protocol goals even when his peer is the Spy. Both protocols appear to achieve their claimed goals.

Chapter 14 concludes the book with a few final remarks, and summarises our contribution through its key concepts. It also briefly advances some lines of possible future work. Finally, it comments on some statistics about file sizes, proof runtimes on two common and inexpensive platforms, and human efforts necessary for the entire book.

There are four appendices to complete the presentation. They present a few relevant fragments of the proof scripts about the main protocols that the book discusses. Such fragments are released with the 2006 distribution of Isabelle [33, 34] (before that distribution appears, they are available with the development snapshot [156]).

Appendix A concerns the Kerberos IV protocol, presenting the guarantees of reliability, session-key compromise, and session-key confidentiality.

Appendix B concerns the Kerberos V protocol, presenting the guarantees of unicity, unicity relying on timestamps, and conjunct key distribution and non-injective agreement.

Appendix C concerns the Shoup-Rubin protocol, presenting the definitions of two important functions and related technical lemmas, and the guarantees of authentication.

Appendix D concerns the Zhou-Gollmann protocol, presenting the guarantees of validity of main and subsidiary evidence, and fairness.

2. The Analysis of Security Protocols

Several approaches may be taken to analysing security protocols formally. Each approach has strengths and limitations, while it is certain that machine support can dramatically assist. The findings may not be straightforward to interpret.

The 1990s increased the awareness that formal methods can significantly contribute to the analysis of security protocols. This decade seems to have turned that awareness into an established fact. In this chapter, we outline a few of the approaches to protocol analysis that stand out for originality (§2.1), and provide some general examples of interpreting their findings (§2.2).

We will see that *Abstract State Machines* (§2.1.1) allow for a formal and rigourous representation of the operational semantics of protocols, although they currently appear to demand deeper integration with mechanical tools. *Belief Logics* (§2.1.2) give a formal representation of the beliefs that the peers derive during the execution of a protocol, but fail to capture several protocol weaknesses. They are a milestone in the field, opening the ground to a number of research efforts. Lately, *Constraint Programming* (§2.1.3) has advanced a new perspective: a quantitative analysis of security protocols. Such analysis proceeds from the observation that each protocol goal is always reached in reality at a certain level (of strength, robustness, reliability, etc.), rather than entirely in the boolean sense. *Provable Security* (§2.1.4) allows for a probabilistic study of the secrecy goals. It is the only notable method that disposes with the black-box assumption on cryptography, and instead relies on a probabilistic account of the robustness of the crypto-algorithms. The *Spi-calculus* (§2.1.5) can effectively reason about the main protocol goals. It extends a popular concurrent language with cryptographic primitives and a formalisation of a malicious attacker. *State Enumeration* (§2.1.6) verifies exhaustively that a protocol model of limited size admits no attacks. Although it records a number of efforts to exceed its limitation to finite models, it has notably advanced the field of protocol verification. Also *Strand Spaces* (§2.1.7) are relevant, as they can be used to elegantly model a variety of detailed protocol features. Their mechanical tool support seems to be developing quickly. The Inductive Method, which is discussed in the next chapter, appears to embody the most desirable features: it can reason about a variety of protocol goals on models of unbounded size, and is mechanically supported by the theorem prover Isabelle.

Interpreting the contributions of protocol analysis is not straightforward. The effort put in interpreting the findings is often unrelated to the effort in obtaining them. We present examples of findings that are easy to obtain and interpret, on the TMN protocol (§2.2.1); fairly easy to obtain but fairly difficult to interpret, on the Woo-Lam protocol (§2.2.2); difficult to obtain but easy to interpret, on the public-key Needham-Schroeder protocol (§2.2.3); and difficult to obtain and interpret, on the shared-key Needham-Schroeder protocol (§2.2.4). The list does not attempt to be exhaustive but merely to highlight that most issues about the protocols and their analyses are not at all simple, although they may, at first, appear to be.

2.1 Formal Approaches

This section outlines some of the most original formal approaches taken thus far for analysing security protocols. It will be clear that each of them has both strengths and limitations. The presentation develops in alphabetical order rather than along the timeline. It has an informative style but no aim to provide a complete survey.

2.1.1 Abstract State Machines

Gurevich's *Abstract State Machines*, ASMs in brief [54], formerly known as *Evolving Algebras* [86], are born as a general-purpose formalism that should be more flexible than a Turing machine but retain the same power. Therefore, the ASM thesis is that any computable program can be represented by a suitable ASM. An ASM is essentially a first-order signature equipped with a program that is a set of if-then-else rules. Once an interpretation of the signature is provided, a static algebra originates, the *initial state* of the ASM. Such a state can be updated by the rules of the program, which, updating certain elements of the signature, produce a new algebra, namely a new *state*. The admissible functions include *oracles*, whose interpretation is not influenced by the ASM program, but is provided by the environment at each state. In consequence, the resulting computational model is not linear but has a graph structure.

The formalism, which has been benchmarked on a large variety of real-world applications [53], has also a distributed variant where each agent can independently run his own program. We tailored this variant for the analysis of security protocols. Initially, we modelled the Kerberos IV protocol using stepwise refinements in the presence of the Spy. Confidentiality at the more detailed level was investigated by simulation [45]. This was the first formal specification of the entire protocol, obtained from the substantial informal documentation provided by its designers [121]. It has significantly simplified the task of deriving an inductive model for the protocol, which we present below along with its correctness proofs (Chapter 7). Then, we developed a general theory of messages and demonstrated it on the public-key Needham-Schroeder protocol [46]. Proofs were conducted by induction but still by pen and paper. However, integration with mechanised tools, such as model checkers [165] or theorem provers [80], has now reached an advanced stage of development.

Rosenzweig et al. take an orthogonal approach [140] adopting ASML [154], a programming language for ASMs recently developed at Microsoft. Specifications written in ASML are therefore fully executable and interoperable with other languages of the .NET family. As the language is object-oriented, basic items such as cryptographic keys and related operations can be easily described as classes extending other basic classes. The resulting treatment is neat and fairly general, arriving at an abstract formalisation of confidentiality in terms of indistinguishability of messages. The machinery is demonstrated on simple protocols, and properties are established by mathematical proofs. Applications to real-world protocols are expected.

2.1.2 Belief Logics

The belief logic due to Burrows et al. [58] counts as the first significant attempt to exceed the limits of informal reasoning. The main idea is to formally represent the beliefs that the agents running a protocol derive at the various stages of the execution. Each step of the protocol is idealised as an (initial) logical formula, while a set of logical postulates is provided. The formulae that, using the postulates, can be derived from the initial ones formalise the goals of the protocol. For the sake of demonstration, we present three typical formulae with their respective semantics [58, p. 236].

– "P would be entitled to believe X. In particular, the principal P may act as though X is true."

$$P \models X$$

It is often denoted also as P believes X.

– "P and Q may use the shared key K to communicate. The key K is good, in that it will never be discovered by any principal except P or Q, or a principal trusted by either P or Q."

$$P \xleftrightarrow{K} Q$$

It is often denoted also as P and Q share K.

– "The formula X is fresh, that is, X has not been sent inside a message at any time before the current run of the protocol."

$$\#(X)$$

It is often denoted also as X is_fresh.

The semantics provided by the authors turned out to be unsatisfactory and several attempts were made to repair the problem [8, 52]. For example, it is not clear from the first formula if P might believe something that is in fact false. The BAN logic indeed claims correct some protocols that were subsequently found to be flawed, such as a variant of the Otway-Rees protocol (Figure 8.4) that does not encrypt the nonce issued by the protocol responder. An attack is possible, whereby the protocol initiator would share a session key with the Spy at completion of the protocol, while believing to be sharing it with the intended responder [132, §3.8]. This is both a violation of confidentiality and of authentication. Another example is the public-key Needham-Schroeder protocol (Figure 2.5), which passed the logic analysis without inconvenience: Lowe discovered a few years later that the protocol suffers a subtle failure of authentication [106].

The second formula has been considered ambiguous because it embodies the goals of both confidentiality and key distribution. The third formula is crucial to the analyses. For example, it is used to specify that the shared-key Needham-Schroeder protocol (Figure 2.6) does not guarantee the protocol responder that the received session key is fresh. In fact, Denning and Sacco had already pointed out that the Spy might fool the responder into accepting an expired session key as a fresh one [71].

A major limitation of the belief logics is how to reason about confidentiality. This is done informally on top of the formulae derived through the calculus of the logic, since no malicious entity is modelled explicitly. A number of extensions have been designed in order to enhance the expressiveness of the logic and account for further protocol goals [3, 82, 115] but they tend to sacrifice the intuitive nature of the logic itself. Proofs by belief logics are typically short and carried out by hand, but certain logics have been implemented [56, 57] using the theorem prover HOL [85].

Cervesato et al. [60] have recently developed yet another logic to study authentication but not confidentiality. Its main merit is a clear identification of the relation between the two goals: they often assume each other. The logic clearly states those confidentiality assumptions that are needed to reason about authentication. Other methods can then be used to establish whether those assumptions are met.

2.1.3 Constraint Programming

This method specifies a system in a declarative style through a number of constraints that the system must satisfy. A *crisp* constraint is a predicate over certain variables and corresponding values taken from a domain set. If the constraint is satisfied, then the variables must take those values. A *soft* constraint is a function that associates domain values with variables at a certain level, which can be variously interpreted, taken from a partially ordered set [49]. Once a system is specified as a constraint satisfaction problem, studying the system reduces to solving the corresponding satisfaction problem, which can be efficiently done algorithmically.

Constraint programming has provided important support to other methods of analysis based on state enumeration, which we mention below (§2.1.6). These establish security properties such as confidentiality when a finite number of possible states of the system under analysis can be checked to maintain the properties. This *reachability* problem is undecidable in the general case [76], but Millen and Shmatikov show that it can be decided in a simplified case by reducing it to a constraint solving problem [120]. Later, Corin and Etalle build up on this work [65].

Constraint programming has also helped us study in detail security problems such as coherence of integrity policies [50] and confidentiality achieved by security protocols [29]. In particular, the use of soft (rather than crisp) constraints allows for a quantitative analysis of the target system [31]. The formal properties are no longer mere yes-or-no predicates but carry an extra parameter expressing their strength or, broadly speaking, their security. For example, a crisp analysis may find out that an implementation of an integrity policy does not uphold some stated requirements. The analyser might move on to a soft analysis, looking for the best of the previously unacceptable implementations.

The same analysis performed over the network scenarios induced by a security protocol can, for example, distinguish more confidential messages from less confidential ones. Formal statements such as "this message component has maximum confidentiality" or "this message has unacceptable confidentiality" are possible because each property carries the mentioned security level. Because of the declarative style of the specification, the method has been stretched to authentication properties with some difficulty, although the final outcome [30] is surprisingly expressive.

2.1.4 Provable Security

Provable security is a complexity-theoretic study of confidentiality. Originally developed by Bellare and Rogaway to address the problem of two-agent authenticated key exchange [47], the notion was later extended by the same authors to cope with the three-agent setting where a trusted Server helps to achieve the goal of distributing session keys to a pair of peers [48]. They comment that devoting the entire effort of formal analysis towards establishing that a protocol suffers no attacks is unsatisfactory: "there is finally a general consensus that session key distribution is not a goal adequately addressed by giving a protocol for which the authors can find no attacks" [48, §1.2].

In this setting, Bellare and Rogaway formally define the problem of session key distribution, design a security protocol and prove it *secure* assuming the existence of a family of pseudorandom functions (PRFs). To show that this assumption is minimal, they prove that if a secure session key distribution protocol exists, then a one-way function exists; then they apply the existing result stating that a PRF family exists if a one-way function exists [90]. The protocol is claimed secure in terms of two properties. The first is key distribution, signifying that both peers know the same session key at completion of the session. The second is the Spy's probabilistically negligible advantage on the discovery of the key, namely confidentiality of the key. All relevant proofs are carried out by hand and involve substantial formal overhead. But this work is a fine piece of theoretical research.

Shoup and Rubin later extend the approach with an account of smartcards, design a new protocol based on smartcards, and prove it secure. It would seem that provable security cannot express certain details of the confidentiality goal which we discuss below (Chapter 11) in relation to a clear statement of the underlying threat model. The actual protocol design must be adapted for the analysis, rather than vice versa, a practice that may raise the risks of verifying a different design. Shoup and Rubin seem to support this view, as they state that "several modifications to the protocol were necessary to obtain our proof of security, even though it is not clear that without these modifications the protocol is insecure" [148, §2.2]. Along the same lines, the implementors of the Shoup-Rubin protocol also point out that "the details of Shoup-Rubin are fairly intricate, in part to satisfy the requirements of an underlying complexity-theoretic framework" [96].

2.1.5 Spi-calculus

The spi-calculus [6] is an extension of the more popular π-calculus [122] with primitives representing the cryptographic operations of encryption and decryption. Like other process calculi, it is based on processes that communicate through channels. Channels may be *restricted* in the sense that only certain processes may communicate on them. The π-calculus and the derived spi-calculus allow the scope of each such restriction to dynamically change during the computation. A process may decide to send a message on a restricted channel to a process outside the scope of the restriction. When this happens, the scope is said to *extrude*.

It is intuitive to use the restricted channels to model confidentiality and the scope extrusion to allow for communication of secrets. Once the Spy is modelled as an arbitrary environment for the protocol, confidentiality of a message X derives from *equivalence checking* two protocol specifications, one featuring X, another an arbitrary X'. The idea of the proof is that the presence of X does not influence the reaction of the environment. However, the details may be difficult to grasp: "If you get lost in the formal passages of the paper, the cleartext nearby may help — hopefully the informal explanations convey the gist of what is being accomplished" [6].

More recent verification strategies based on Woo and Lam's concept of *correspondence assertion* [167] aim at establishing strong causal relations between the protocol events to reason about authentication. Our formalisation of the authentication goal (§4.6), which in general establishes that an event precedes another, seems to have much in common with correspondence assertions, although we do not need to state any size limits. Gordon and Jeffrey initially show how to embed correspondence assertions in the π-calculus and prove them by type checking [84], so that infinite states can be treated. However, the method does not address security issues and hence is only adequate for communication protocols. The same researchers, inspired to Abadi's seminal work on secrecy by typing [1], later extend their technique to the context of the spi-calculus [83]. Properties of security protocols can be treated in the extended framework, which formalises an opponent.

2.1.6 State Enumeration

The well-known process calculus CSP [92] has had vast applications in the field of formal methods thanks to its intuitive notions of *process* and *channel*. This setting easily scales up to the analysis of security protocols [144], as pioneered by Ryan [141]. The approach is mainly targeted at detecting possible attacks by the Spy.

The peers are modelled as processes that exchange the protocol messages via specific channels, thus formally producing new *states* of the model. This first specification accounts for no malicious entity, and hence is certainly not flawed because no one tries to mount attacks. Then, another specification is obtained by introducing the Spy as a new process that can perform illegal operations. A long established proof method is *trace equivalence*. The protocol is claimed to suffer no attacks if the two specifications are trace equivalent, namely if they both reach the same states. Checking this by pen and paper is long and tedious, so a model checker can be tailored to enumerate the reachable states. There are intrinsic limitations: only finite systems of reasonably small size can be tackled. They typically account for at most three or four agents, including the Spy. But various recent techniques such as symbolic model checking [116] have been developed throughout the years as an attempt to overcome this limitation. More recently, the concept of correspondence assertion [167] has also been used in this context, especially to prove authentication properties [142]. Another proof method relies on *rank functions* [142].

Lowe has had successful results with the model checker FDR [139] and with an abstract language called Casper [110] that can be easily compiled into CSP. He has discovered a number of attacks [108, 112]. Other model checkers such as ASTRAL [68], Brutus [62, 63], Murphi [74, 123] and SPIN [19] to just mention a few, have been tailored to protocol verification obtaining perhaps less seminal findings. Recent work by Basin et al. [20] must be mentioned here, as it makes significant improvements over traditional model checking techniques. Infinite elements are represented using *lazy datatypes*: constructors that build datatypes without evaluating their arguments. The potentially infinite messages that the Spy can introduce are treated using a dedicated symbolic representation. The resulting method is tested on a number of classical protocols and on simplified versions of a few real protocols [21]. Also SAT-based techniques have lately boosted the efficiency of model checkers for protocol analysis [11, 16]. In terms of specification, the new techniques often recast the protocol models as planning problems. In terms of verification, they borrow algorithms previously developed for the problem of boolean satisfiability.

In conclusion, if we look at research spanning nearly two decades now, it seems fair to state that the combined use of state enumeration with model checking has significantly contributed to advancing the entire protocol verification field. However, if a system of limited size does not suffer any attacks, it is not in general obvious that neither does a system of arbitrary size. This result has been proved *ad hoc* for a specific protocol by pen and paper [111], while another tool, the NRL Protocol Analyzer [118], also allows mechanised proofs by induction that certain states are out of reach.

2.1.7 Strand Spaces

Another important approach [79] rests on the notion of *strand*: a record of a protocol history from the viewpoint of a single peer. Therefore, a strand is the sequence of events (message sending or receiving) concerning a peer of a protocol. This differs from our notion of *trace*, given in the next chapter, which records an entire protocol history from the viewpoint of an external observer, and therefore features events pertaining to a variety of agents. A *strand space* is an unspecified set of strands, some for the agents of the network, some for the Spy. As can be expected, while the strands formalising the Spy's illegal behaviour are independent from the protocol being specified, those for the protocol peers are not.

The notions of fresh or unguessable components are elegantly modelled as constraints on the construction of strands. For example, there exists no strand in which the Spy sends an unguessable component prior to its reception. More important for verification purposes is the notion of *bundle*, a set of traces that are sufficiently expressive to formalise a protocol session. Therefore, certain strands of a bundle send messages, and other strands receive them. These must be bundled with the Spy's strands to form a new bundle modelling a realistic scenario where the protocol messages can be overheard and prevented from reaching their intended recipients. Proofs are carried out by induction on a bundle and their philosophy is reasonably easy to grasp. The treatment is originally carried out by pen and paper and applied to three classical protocols: Needham-Schroeder [79], Otway-Rees and Yahalom [78].

Studying the authentication goal within this context produces the concept of *authentication test* [77]. It is a formalisation of the simple authentication strategy that sees an agent issue a fresh value and send it over the network. Freshness here assures that the probability that other agents know the value without somehow receiving it from the network is negligible. Therefore, should the sender ever get that value back, he would authenticate the context with which he associated it in the first place. This test is used to study what form of authentication a large class of protocols attains. The strand-space approach is also tried out for the computational analysis of security protocols over the Diffie-Hellmann key-agreement scheme [91] and, more recently, for reasoning about the trust assumptions that the protocol participants silently make [89]. Strands are annotated with formulae of a simple logic that formalises trust management, resulting in the first promising initiative to conjugate trust management and protocol analysis. Temporal logic can assist with the verification methodology [59], while machine support currently is under development [88, 124].

2.2 Interpreting the Findings

We discuss some findings derived from the formal analyses of well-known protocols. Some of the corresponding formal notation is quoted for the sake of demonstration. The only exception is the discussion about the Woo-Lam protocol: it arises from informal reasoning, which remains a useful approach to protocol analysis.

2.2.1 TMN

Let us consider the TMN protocol [153], which aims at distributing session keys for mobile communications (Figure 2.1).

$$
\begin{array}{llllll}
1. & A & \rightarrow & S & : & A, S, B, e(K_A) \\
2. & S & \rightarrow & B & : & S, B, A \\
3. & B & \rightarrow & S & : & B, S, A, e(K_B) \\
4. & S & \rightarrow & A & : & S, A, B, v(K_A, K_B)
\end{array}
$$

Fig. 2.1. TMN protocol

The identifiers A, B, K_A, K_B, abd S respectively represent the protocol initiator, the responder, the keys they choose and the trusted Server; the unary constant $e()$ denotes a standard encryption function that only the Server is able to invert, and the binary constant $v()$ stands for bit-wise exclusive-or (Vernam encryption).

Nothing protects the messages, so, while intercepting them, the Spy can read and modify any components. The protocol fails to enforce agent authentication. The Spy may send the Server a fake instance of the first message using a key chosen by herself rather than by A, and complete the protocol with B. Even if A has not taken part in the session with B, both B and

$\langle fake.Msg1.A.S.B.Encrypt.K_C,$

$comm.Msg2.S.B.A,$

$comm.Msg3.B.S.A.Encrypt.K_B,$

$intercept.Msg4.S.B.A.Vernam.K_C.K_B,$

$resp\_fake\_session.A.B.K_B\rangle$

Fig. 2.2. Authentication attack on TMN protocol: CSP notation

the Server will believe that the opposite is true at completion of the session. Precisely, B will rely on the key K_B for communicating with A.

This scenario [112, Attack 4.1], which can be spotted also while reasoning informally, is fairly easy to interpret. Its formalisation as a list of CSP events is demonstrated in Figure 2.2 and seems intuitive. It sees the Spy send the first message on the channel *fake*, while the second and third messages are issued legally and travel on the channel *comm*. Finally, the Spy prevents the delivery of the fourth message to A on the channel *intercept*, while B erroneously concludes that he shares the key K_B with A.

Similarly, the Spy may intercept the second message, replace B's identity with hers in the third, and include in it a key chosen by herself rather than by B [112, Attack 4.2]. As a result, both A and the Server would believe that B took part in the session, and A would erroneously think that the key invented by the Spy is shared with B.

2.2.2 Woo-Lam

The Woo-Lam protocol [166] aims at authenticating the initiator to the responder (Figure 2.3). According to a standard notation, Ka in general indicates A's long-term key shared with the Server, while Nb stands for the nonce issued by B. Authentication here merely means active presence in the network.

$$
\begin{array}{lllll}
1. & A & \rightarrow & B & : & A \\
2. & B & \rightarrow & A & : & Nb \\
3. & A & \rightarrow & B & : & \{Nb\}_{Ka} \\
4. & B & \rightarrow & S & : & \{A, \{Nb\}_{Ka}\}_{Kb} \\
5. & S & \rightarrow & B & : & \{Nb\}_{Kb}
\end{array}
$$

Fig. 2.3. Woo-Lam protocol

After B receives, encrypted, the nonce that he issued for A, he forwards it to the Server quoting A's identity. The Server extracts the nonce, ultimately using A's shared key, and returns it to B. This convinces B that the nonce was encrypted using A's key, which hence implies A's presence. The protocol does not attempt to convince B that A indeed meant to communicate with him, which would be a stronger form of authentication (§4.6).

$$
\begin{array}{llllll}
1. & C & \rightarrow & B & : & A \\
1'. & C & \rightarrow & B & : & C \\
2. & B & \rightarrow & A & : & Nb \\
2'. & B & \rightarrow & C & : & Nb' \\
3. & C & \rightarrow & B & : & \{Nb\}_{Kc} \\
3'. & C & \rightarrow & B & : & \{Nb\}_{Kc} \\
4. & B & \rightarrow & S & : & \{A, \{Nb\}_{Kc}\}_{Kb} \\
4'. & B & \rightarrow & S & : & \{C, \{Nb\}_{Kc}\}_{Kb} \\
5. & S & \rightarrow & B & : & \{Nb''\}_{Kb} \\
5'. & S & \rightarrow & B & : & \{Nb\}_{Kb} \\
\end{array}
$$

Fig. 2.4. Authentication attack on Woo-Lam protocol

Abadi and Needham point out an attack whereby the Spy impersonates A when communicating with B [7, §4], which implies that B's informal reasoning given above is flawed. The Spy (indicated by C in Figure 2.4) may interleave two sessions with B. She uses A's identity in the first, but acts on her own behalf during the second session (distinguished by the primes). During this session, the Spy cheats in the third step using the nonce Nb that B addressed to A. Therefore, the Server's reply for the first session contains a new nonce Nb'', while the reply for the second session must contain Nb. At the end, B is misled into believing that he has communicated with A.

However, one may wonder why B is not puzzled by receiving on a session the nonce that he issued on another session. It must be stressed that no agent can distinguish to which sessions the received messages belong, unless the contents of the messages state this. Therefore, receiving the nonce previously issued for A gives B the required "evidence" of A's presence although this is a wrong conclusion.

2.2.3 Public-key Needham-Schroeder

Lowe's "middle-person attack" [107] on the public-key Needham-Schroeder protocol [126] is a rather subtle one. It took one and a half decades to be discovered. The protocol can be found in Figure 2.5. In brief, it exchanges the nonces Na and Nb in order to mutually authenticate A and B.

Since A receives in the second step the nonce that she encrypted using B's public key, she decides that B was alive. Upon reception of his own nonce in the third step, B draws the same conclusion about A. Lowe employs model checking techniques to show how the Spy can exploit two interleaved runs and interpose between the peers so that B believes that his peer is A when it

$$1. \quad A \;\to\; B \;:\; \{A, Na\}_{Kb}$$
$$2. \quad B \;\to\; A \;:\; \{Nb, Na\}_{Ka}$$
$$3. \quad A \;\to\; B \;:\; \{Nb\}_{Kb}$$

Fig. 2.5. Public-key Needham-Schroeder protocol

in fact is the Spy [106]. The details of the attack are omitted here as they are among the most published pieces of research in the field of computer security.

Interpreting the attack may not be trivial. It does not violate authentication in terms of aliveness (it is true that A is active in the network, as B believes), but a stronger form of authentication, *weak agreement* (§4.6), whereby B should be assured that A intends to communicate with him. The publication of this attack convinced many that formal methods might be an effective tool to detect the subtle consequences of session interleaving. Moreover, by showing that the attack can be avoided simply by including B's identity in the second step, Lowe points out the importance of explicitness to protocol messages, later investigated more deeply by Abadi and Needham [7].

Paulson formally verifies [133, §5.4] that, if B sends the second message to A and someone sends an instance of the third, then B is assured that A has sent an instance of the first message *to someone*. This is one of the strongest relevant guarantees of authentication of A with B (§5.4). However, it only conveys A's aliveness from B's viewpoint, and fails to convey weak agreement of A with B. It in fact expresses no evidence about A's putative wish to communicate with B. The peers' viewpoints are crucial to realistic guarantees, as we anticipate in the next section and treat extensively later (Chapter 5).

2.2.4 Shared-key Needham-Schroeder

Needham and Schroeder also designed a key distribution protocol based on symmetric encryption [126]. It presupposes that each agent shares a long-term key with the trusted Server, and is presented in Figure 2.6.

$$1. \quad A \;\to\; S \;:\; A, B, Na$$
$$2. \quad S \;\to\; A \;:\; \{Na, B, Kab, \{Kab, A\}_{Kb}\}_{Ka}$$
$$3. \quad A \;\to\; B \;:\; \{Kab, A\}_{Kb}$$
$$4. \quad B \;\to\; A \;:\; \{Nb\}_{Kab}$$
$$5. \quad A \;\to\; B \;:\; \{Nb - 1\}_{Kab}$$

Fig. 2.6. Shared-key Needham-Schroeder protocol

A replay attack on the protocol is well known. The Spy may intercept the certificate sent in the third step and later replay it to B without understanding its contents. As a result, B may be fooled into accepting an old session key as fresh.

Paulson proves an important theorem by induction: the protocol guarantees confidentiality of the session key issued by the Server, provided that both peers' shared keys are safe from the Spy [33, 34]. We anticipate some formal syntax here, as we find it mostly self-explanatory; it will be detailed later (Chapter 3). The main assumption of the theorem requires that the event

$$\text{Says Server } A \, (\text{Crypt}(\text{shrK } A)\{\!\mid\! Na, \text{Agent } B, \text{Key } Kab, X \!\mid\!\}) \tag{2.1}$$

occur. However, the theorem must be interpreted with care. Can the peers who are intended to use the session key ever appeal to it? In general, no agent can verify any events concerning other agents because they take place at other points in the network. Therefore, the theorem is only applicable by the Server or by some super-agent who sees all occurring events. It is not applicable either by A or by B, who cannot inspect the Server's activity unless we establish some suitable lemmas on assumptions that they can verify. In particular, we can prove that, if A's shared key is safe from the Spy and the certificate

$$\text{Crypt}(\text{shrK } A)\{\!\mid\! Na, \text{Agent } B, \text{Key } Kab, X \!\mid\!\}$$

appears in the traffic, then the event 2.1 occurred. Then, Kab is confidential by application of the previous theorem. The resulting theorem can be applied by A because, if she ever receives an instance of the second message, the message was in the traffic.

Similarly, the confidentiality argument can be made useful to B by the following lemma. If B's shared key is safe from the Spy and the certificate

$$\text{Crypt}(\text{shrK } B)\{\!\mid\! \text{Key } Kab, \text{Agent } A \!\mid\!\}$$

appears in the traffic, then the event 2.1 occurred for some Na. Consequently, when B receives an instance of the third message, he considers the session key found inside it to be confidential.

Following these considerations, we emphasise that it is the peers running the protocol who need guarantees that the protocol goals are met. On this basis, we will develop the principle of *goal availability* for formal protocol analysis (Chapter 5). In short, it says that any formal reasoning must be conducted in terms of assumptions that the protocol participants can verify. This pays back in terms of protocol insights that would remain hidden otherwise.

3. The Inductive Method

Formal correctness of security protocols can be established by means of the Inductive Method with mechanical support from the theorem prover Isabelle. The protocol goals can be studied over an inductively defined model using the corresponding induction principle.

The informal way of studying a protocol property is to somehow derive persuasive reasoning that all protocol steps preserve the property. If not complex *per se*, the process tends to reach unmanageable size because a protocol is a concurrent program that is executed by an indefinitely large population of agents. It was exactly the aim of preserving a property through various steps that inspired the Inductive Method [133]: proving the property formally via structural induction on an unbounded protocol model.

Let us consider a security protocol, $\mathcal{P}$, and an unlimited population of agents. To account for some threat model, the agents include a Spy who monitors the entire network. The Spy also knows the long-term secrets of an unspecified set of compromised agents, a feature that models some form of collusion. The network traffic develops according to the actions performed by the agents while they are executing various sessions of $\mathcal{P}$. Each agent can interleave an unlimited number of protocol sessions. A history of the network traffic can be represented by the list of the events that occurred throughout that history. Such a list is a *trace*, and is typically built in reverse chronological order. The set P of all possible traces of unbounded length is a formal model for the network where $\mathcal{P}$ is executed. By misuse of terminology, P is referred to as the *formal protocol model for* $\mathcal{P}$. The set P is defined inductively by specific rules drawn from $\mathcal{P}$. The events occur via the firing of these rules. Since no rule is forced to fire, no event is forced to occur.

The use of induction in analysing security protocols is also present in the work of Meadows [118], who combines state enumeration on a system of limited size with inductive proofs that the system never reaches infinite sets of states. However, the Inductive Method [133] is entirely inductive. Induction during the specification phase provides a way to define all relevant message operators and the protocol model itself. During the verification phase it provides the proof principle to establish properties of the protocol model. So, we only verify safety properties; we cannot prove liveness properties, namely that something will occur. The resulting operational semantics has much in common with CSP formalisations [142], except that its models are unbounded. The approach is mechanised by the generic theorem prover *Isabelle* [130], which offers strong support for inductively defined sets, efficient simplification by conditional rewrite rules and automatic case splits for if-then-else expressions.

This chapter briefly introduces Isabelle (§3.1) with its theory hierarchy (§3.2), and then reviews the original version of the Inductive Method prior to our extensions. We will see a formalisation of the agents running the protocols (§3.3), of the cryptographic keys (§3.4) and of the compromised agents (§3.5). Then, we will introduce a formal treatment of the protocol messages (§3.6), of the protocol events (§3.7) and of the traces of events (§3.8). We will terminate the chapter with an account of the threat model in which the protocols are imagined to be executed (§3.9), of the main operators necessary for the formal treatment (§3.10) and, finally, of the protocol model (§3.11).

3.1 Isabelle

Isabelle is a generic, interactive theorem prover [130]. *Generic* means that it can reason in a variety of formal systems. This book refers to the best developed and most popular version, Isabelle/HOL [128]; it supports *higher-order logic*, a typed formalism that allows quantification over functions, predicates and sets. Hardware and software systems can readily be modelled in higher-order logic, and correctness properties expressed. *Interactive* means that it is not entirely automatic and, rather, requires a good amount of human intervention. But Isabelle also provides much automation. Its *simplifier*, which can be invoked by the proof method `simp`, combines rewriting with arithmetic decision procedures. Its *automatic provers*, corresponding to various proof methods, such as `auto`, `blast` and `force`, can solve most simple proof scenarios.

Most proofs are conducted interactively. In a typical proof, the user directs Isabelle to perform a certain induction and then to simplify the resulting subgoals. Any surviving subgoals might be given to an automatic prover or be reduced to other subgoals by the use of some lemma. Failure to find a proof for a conjecture may simply mean that the user is not skilled enough; otherwise, it may exhibit what in the system being modelled contradicts the conjecture and hence help in locating a system bug.

The series of commands used to prove a theorem can be seen as a proof sketch. Confidence that the proof is sound comes from observing the line of reasoning that we are forced to adopt, and the lemmas we are forced to prove. Interactive theorem proving is difficult, and this very difficulty strengthens our confidence that the resulting theorems are true. Conversely, a fully automatic proof may lead to worries that the model of our system is too abstract. When a theorem has been proved, Isabelle can deliver a formal proof object; as automation improves, people will increasingly want to examine these proof objects (perhaps using an independent tool) in order to assure themselves that the proof is valid.

Security protocols can be modelled using general-purpose tools such as Isabelle. Once the models have been shown to be useful, researchers will naturally use them as the basis for specialised protocol verification tools, which may achieve high performance. Two eminent examples of specialised tools are Cohen's TAPS [64], an automatic theorem prover based on first-order logic, and Blanchet's ProVerif [51], based on declarative programming. Isabelle remains useful for modelling novel protocols that are not covered by the specialised tools.

Isabelle can be freely downloaded from the Internet [157] under the Open Source Software BSD licence. It is available for most Unix platforms, such as Linux, MacOS X and Solaris. For Microsoft Windows platforms, it is sufficient to install a Linux-like environment such as Cygwin (which comes under the GNU General Public Licence) [155]. Installing Isabelle reduces to unpacking the relevant files, which include an ML compiler and runtime system, such as Poly/ML [159] or Standard ML of New Jersey [161], and the latest version of the Proof General graphical interface [160].

Each distribution of Isabelle comes with a repository of proofs. A new distribution is released every year to include the latest developments to the provers and optimisations to the proof scripts installed in the repository. This is the main reason why this book concentrates on proof concepts and methods rather than on proof scripts, which are limited to a few examples throughout the text and some demonstration in the appendices. While scripts are certain to change over time, methods are not.

All proof scripts about the protocols treated in this book can be found in the local repository [33] or in the online repository [34] from the 2006 distribution. Before that distribution is released, they are available with the development snapshot [156]. A profitable way to read this book over the years is to interactively execute the current proof scripts from the Isabelle repository while the theorems are presented throughout. The task will be simplified by total uniformity of theorem names (§1.3.2) between the book and the proof scripts. It will demonstrate step by step the mechanised support for our considerations.

3.2 Theory Hierarchy

The process of studying a system begins with the development of an Isabelle specification of the system. Such a specification is called *theory*. An Isabelle theory also contains the theorems that are found to hold for the specification, and their proof scripts. Each theorem statement is followed by its proof script. The user may also decide to retain failed proof attempts that are particularly instructive. For example, the theory file List.thy contains a simple specification of the datatype of finite lists, and the definitions of a number of functions to manipulate lists, such as butlast and concat. The same file contains theorems expressing various properties of those functions.

Theories can be variously developed on top of each other to form an effective theory hierarchy. Each theory inherits all structures and theorems that were developed within its ancestor theories. Figure 3.1 presents the theory hierarchy that is relevant to this book. It is a fragment (subtree) of the entire theory hierarchy that comes with the Isabelle repository [33, 34]. It was automatically generated using the isatool command [128].

The theory file Message.thy contains a formal account of the messages that security protocols exchange, including the main operators to reason about them. It is a child theory of the HOL database of theories, which itself is a child of the Pure database. One step down in the hierarchy, we find the theory file Event.thy, which accounts for the network events, such as sending a protocol message. Another step down is the file Public.thy, which contains a formalisation of both symmetric and asymmetric cryptography. It is useful for protocols that use both techniques, such as the accountability protocols (Chapter 13); its earlier distributions contained a treatment of asymmetric cryptography only. These three files terminate the mechanisation of the core components of the Inductive Method in Isabelle. Observe, however, that the method itself is technically independent of Isabelle. It could be potentially mechanised on any mechanical prover or used by pen and paper, though in the latter case the limits of human intuition would constrain the findings.

The theory file for a protocol bears the name of the protocol or its designers. The protocol theories all lie at the same leaf level in the hierarchy. They include two with the _Gets suffix, indicating that the network event of receiving a message is included, in contrast with the corresponding theory names without that suffix, where the reception event is not present. These theories can be siblings because the current version of Event.thy accounts for message reception following our developments (§8.2), whereas reception was not originally available. When there is only one file version, the file itself indicates whether it includes reception.

It can be observed that, while the theory files ZhouGollmann.thy and CertifiedEmail.thy for the accountability protocols follow the conventional hierarchy, smartcard protocols cause a branching point. Theory EventSC.thy contains an extended set of events to formalise protocol interaction with smartcards, and it is therefore a sibling of Event.thy. Its child theory named Smartcard.thy completes the formal account of the cards with a specification of their stored secrets and of the threats of theft or cloning. The leaf theories ShoupRubin.thy and ShoupRubinBella.thy contain the specification and verification of two versions of a smartcard protocol.

As mentioned in the previous section, this book refers to Isabelle/HOL, which implements the higher-order logic formalism. The earlier Isabelle syntax was much in the style of the ML programming language. The recent *Isar* syntax [164] allows for a more natural interaction with the prover through a

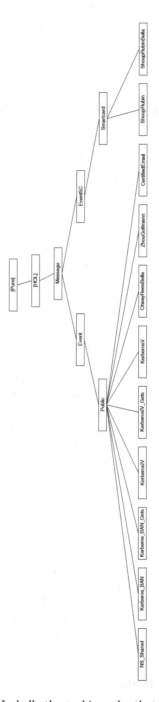

Fig. 3.1. Fragment of Isabelle theory hierarchy that is relevant to this book

more readable and intuitive syntax. Most of its constructs seem to be self-explanatory. For example, it retains the obvious construct for a datatype, which is of great use with security protocols as it provides a compact and easily extensible way to define type constructors that are injective and have disjoint ranges. As we shall see, a datatype can be used to define types agent, msg and event for respective, crucial elements. Their formal treatment, along with that for cryptographic keys and for a Spy implementing a threat model, is conveniently spread across the three main files Message.thy, Event.thy and Public.thy. However, we prefer to present it more organically and didactically below, without adherence to the specific theory file but, rather, with adherence to each specific element.

3.3 Agents

Although finite, the population of agents running a security protocol in the real world is indefinitely large. Therefore, modelling any limited population would prevent the protocol model from capturing some potentially realistic scenarios.

An infinite population can be easily modelled by establishing a bijection with the set of natural numbers. Given a number i, we indicate the corresponding agent as Friend i. A malicious agent can be modelled by the nullary constructor Spy, which will be used to formalise a realistic threat model (§3.9) against which to investigate robustness of a protocol. Most symmetric-cryptography protocols rely on a trusted third party, the Server, which has access to all agents' long-term secrets. It is modelled by Server and often abbreviated as S. These agents can be compactly introduced as

datatype agent $\triangleq$ Server

Friend nat

Spy

In particular, Kerberos IV (Chapter 7) and V (Chapter 9) have two trusted parties, so the datatype should be extended accordingly, though we will find a simple heuristic to avoid it (§7.2). "Agents could be interpreted as humans, machines, or processes" [7, §2]. We prefer the last interpretation as we do not model the computer architecture between processes and humans. If a confidentiality guarantee is available to a process, then it is not necessarily available to the human owner of the process because the Spy could break in at any architectural level and invalidate the guarantee before it reaches the human.

3.4 Cryptographic Keys

We assume familiarity with the concepts of encryption and digital signature.

The free type key is introduced to formally represent cryptographic keys. In the symmetric-cryptography setting, each agent is endowed with a long-term key that is shared with the Server. It is trivially modelled as

shrK : agent $\longrightarrow$ key

The same declaration applies to pairs of functions modelling asymmetric key pairs for encryption, priEK and pubEK, or signature, priSK and pubSK. When the distinction between the two key pairs is unnecessary, such as with most classical protocols, one pair of functions is translated into priK and pubK, and the other one is ignored.

The set symKeys of symmetric keys helps us distinguish between various keys. In particular, recall that a *session key* is a fresh symmetric key that some protocols aim to distribute to their peers and is only valid for the current session. A session key K can be easily modelled as

$$K \in \text{symKeys} \quad \text{and} \quad K \notin \text{range shrK}$$

This clearly presupposes that the symmetric keys can be partitioned into shared keys and session keys, an assumption met by all experiments conducted thus far. Should other kinds of symmetric keys be on the scene, an alternative formalisation of session keys would be necessary.

Another function is useful below to define the message operators compactly (§3.10). It is

invKey : key $\longrightarrow$ key

and leaves a key unaltered in case it is symmetric; otherwise, it turns the key into its corresponding asymmetric half. For example, priSK $A =$ invKey(pubSK A).

3.5 Compromised Agents

A set of *compromised* agents is introduced. It is an unspecified set bad of agents who have revealed their respective secrets to the Spy since the beginning of the protocol. Secrets comprise their keys shared with the Server and their private keys. Compromised agents have another important feature. They also reveal to the Spy any notes of message components they may take during the protocol sessions. Trusted servers are always assumed to be *uncompromised*, namely not to belong to bad.

It is convenient to state that the Spy belongs to bad. With most protocol guarantees that only hold for an agent other than the Spy, we tend to assume that agent to be uncompromised, which automatically rules out the case in which she is the Spy. This makes the theorem statements generally more uniform throughout the various chapters.

Only with accountability protocols (Chapters 12 and 13) shall we see that reliance on the set of compromised agents is too optimistic. With that group of protocols the peers do not trust each other, and therefore we shall define the subset of compromised agents that exclude the Spy.

3.6 Messages

The original datatype of messages includes six constructors [132, §3.1]. It is

datatype msg $\triangleq$ Agent agent

Nonce nat

Key key

Mpair msg msg

Hash msg

Crypt key msg

The basic constructors allow agent names, nonces and cryptographic keys. Observe that constructor Agent requires a variable of type agent, which itself is a datatype (§3.3), while constructor Nonce takes a natural number. Constructor Key takes a variable of type key, which was introduced above (§3.4) as a free type.

The recursive constructors introduce concatenated messages, hashed messages and ciphers. An n-component concatenated message can be obtained using the MPair constructor. For brevity, it is indicated as $X_1, \ldots X_{n-1}, X_n$, external fat braces being omitted according to the notational conventions seen above (§1.3.1).

The encryption primitive Crypt is used to formalise ciphertexts created by either symmetric or asymmetric encryption and also to represent digital signatures. As is customary, encryption is assumed to be perfect: we will see that there is no rule to extract X from Crypt KX unless K is available (§3.10). Also, encryption is assumed to introduce enough redundancy to be collision-free. This obviously does not hold for certain encryption schemes such as exclusive-or, which would require an alternative formalisation of the datatype. Guessable numbers are introduced below (§6.1) as an extension to the original datatype of messages. They serve to model timestamps.

3.7 Events

The Inductive Method initially allows for only two formal events formalising the acts of sending or noting a message

datatype event $\triangleq$ Says agent agent msg

Notes agent msg

The former requires three parameters: one for the agent who sends the message, one for the agent who is the intended recipient and one for the actual message. The latter requires only two parameters: the agent who notes down the message, and the message itself.

The necessity of the sending event is intuitive, though this may not be the case with the noting event, which is used when agents need to note down portions of messages they receive for subsequent use, as is the case with the analysis of the TLS protocol [134]. Message noting is also used when the Spy learns a session key because of some agent's inaccuracy, as we explain later in this chapter (§3.11). It is of great use to most protocol specifications. Later, we will introduce a third event to model message reception and derive a formalisation of agents' knowledge (§8.2). Additional events will be necessary for modelling protocols based on smartcards (§10.2).

3.8 Traces

Event lists of any length, which are called *traces*, are particularly important. Traces will be the parameter for the construction of the protocol model and for the verification of its properties (§3.11). A trace can be used to record all those events occurring during a history of the network while a protocol is in use, and thus to formalise one possible fate of that protocol.

Says Spy D {|Agent A, Nonce Nc|}

Notes Spy {|Nonce Nc, Key Kcb|}

Notes Spy {|Nonce Nc, Key Kcb|}

Says B C (Crypt (priSK B) {|Nonce Nc, Key Kcb|})

Notes B (Nonce Nc)

Says Spy D (Nonce Na')

Says C B {|Agent C, Nonce Nc|}

Notes Spy (Nonce Na')

Says A B {|Agent A, Nonce Na'|}

Says A B {|Agent A, Nonce Na|}

Fig. 3.2. Example trace

Traces are built in reverse chronological order, as we shall see (§3.11), the most recent event being added at the head. For the sake of demonstration, let us consider the example trace in Figure 3.2. It counts ten events. The first event sees A send her identity and a nonce Na to B, who decides to never reply to A during the very history of the network that the trace models. The second event sees another attempt by A with another nonce Na'. Then, we find an event whereby the Spy overhears a nonce. The fourth event sees C send his identity and a nonce Nc to B. With the fifth event, the Spy illegally sends someone else's nonce, which she noted earlier. After that, an event records B's act of noting down the nonce that C sent him. Then, we have B's reply to C. The Spy notes down the body of this ciphertext, an event that occurs twice. (The protocol model lets an event occur an indefinite number of times. However, if an agent is required to send a fresh nonce, namely one that has just been created, then the corresponding event can only occur once.) Finally, the last event in the trace models another illegal event performed by the Spy.

This trace is built having in mind the example protocol seen above (Figure 1.1). However, it is important to remark that, to establish whether the trace really is admissible in the model for the example protocol, we must see how that model is constructed (§3.11). The protocol model will define all and the only traces that the protocol under study can produce.

The notation of session keys routinely indicates both the peers that each key pertains to, but with nonces a single letter indicating only the originator is preferred when there is no ambiguity. For example, to extend the example trace with an event whereby C sends a fresh nonce to D, that nonce would have to be indicated as Ncd. Likewise, Na would be indicated as Nab, then Nc as Ncb, and Nb as Nbc. We face a similar notational issue when indicating nonces pertaining to the same peers but to different sessions; here, we merely add a prime to nonce Na.

3.9 Threat Model

A realistic analysis of security protocols must account for a realistic threat model, which the protocol goals must withstand. We adopt the standard threat model due to Dolev and Yao [75]. It defines the most hostile environment for the protocols, except for ruling out cryptanalysis. It rests on a powerful Spy agent, and can be summarised as follows.

1. *The Spy is a legitimate agent.* She is a legally registered agent like everyone else; so she is at least capable of legal action. For example, in a symmetric-cryptography setting, we can imagine that the Spy has her own long-term key shared with the trusted Server. In an asymmetric-cryptography setting, she will have her own public key appropriately certified by an authority.
2. *The Spy controls the network traffic.* She gets hold of every message that is sent in the network, and can manipulate it as the next item describes.

Therefore, she can prevent delivery of any message or redirect it to agents other than the intended recipient. In this setting, confidentiality and authentication for example become particularly difficult to enforce.

3. *The Spy can perform any message operation except cryptanalysis.* She can break up messages into components by splitting up concatenated ones and opening up ciphertexts sealed with keys that she knows. Using the learnt components, she can then form new messages at will by concatenation and encryption. However, she cannot use any cryptanalytic technique to derive significant information from ciphertexts of which she ignores the encrypting key. This is equivalent to assuming encryption to be perfect.

We can now move on to describe how the Inductive Method formalises this threat model. In fact, it formalises an extended version by allowing, in addition to the three requirements introduced here, the Spy to collude with the set of compromised agents mentioned above (§3.5).

Requirement 1, that the Spy is a legitimate agent, is met by formalising protocol agents (§3.3) and cryptographic keys (§3.4) as we have seen. It is also important that the protocol model (§3.11) will allow any agent, including the Spy, to execute the rules for each legal protocol step.

Meeting requirement 2 requires a formalisation of the Spy's initial knowledge, namely what she knows before any protocol session takes place. For this purpose, the original formulation of the Inductive Method includes the function

$$\text{initState} : \text{agent} \longrightarrow \text{msg set}$$

It also models the initial knowledge of agents different from the Spy because agent is a datatype (§3.3). However, we will see that before our dedicated study on agents' knowledge (Chapter 8), this was not of any use. To understand the function definition, the reader should recall that ` indicates the image operator.

1. The Server's initial knowledge consists of all shared keys, all public keys and only its own private keys.

   initState Server $\triangleq$ (Key ` range shrK) ∪
   {Key (priEK Server)} ∪ {Key (priSK Server)} ∪
   (Key ` range pubEK) ∪ (Key ` range pubSK)

2. Each friendly agent's initial knowledge consists of his own shared key and private keys, and of all public keys.

   initState (Friend i) $\triangleq$ {Key (shrK (Friend i))} ∪
   {Key (priEK (Friend i))} ∪ {Key (priSK (Friend i))} ∪
   (Key ` range pubEK) ∪ (Key ` range pubSK)

3. The Spy's initial knowledge consists of all compromised agents' shared keys and private keys, and of all public keys. Since the Spy is assumed to be compromised (§3.5), by this definition she knows her own keys.

$$\mathsf{initState\,Spy} \;\triangleq\; (\mathsf{Key\,`\,shrK\,`\,bad}) \cup$$
$$(\mathsf{Key\,`\,priEK\,`\,bad}) \cup (\mathsf{Key\,`\,priSK\,`\,bad}) \cup$$
$$(\mathsf{Key\,`\,range\,pubEK}) \cup (\mathsf{Key\,`\,range\,pubSK})$$

This terminates the formalisation of the static knowledge that agents possess before engaging in any protocols, but we still have not met requirement 2 of the implementation of Dolev-Yao's threat model.

No formalisation of agents' dynamic knowledge was available in the original version of the Inductive Method; we shall formalise it later (Chapter 8). However, only the Spy's dynamic knowledge, which she obtains from inspecting protocol traces, was formalised. In brief, we require the Spy to know her initial state, any message ever sent by anyone and any message ever noted by a compromised agent. The function

$$\mathsf{spies} \;:\; \mathsf{event\ list} \longrightarrow \mathsf{msg\ set}$$

formalises the Spy's dynamic knowledge acquired from a protocol trace. Recall that $\#$ formalises the list cons operator, hence $\mathsf{Says}\,A\,B\,X \# evs$ for example is the trace whose first event is A's sending X to B and the rest is a subtrace evs. The function spies can be easily defined by primitive recursion as follows.

0. The Spy's initial state is known to her on any trace, including the empty one.

$$\mathsf{spies}\,[] \;\triangleq\; \mathsf{initState\,Spy}$$

1. Any message sent by anyone in a trace is known to the Spy on that trace.

$$\mathsf{spies}\,((\mathsf{Says}\,A\,B\,X)\,\#\,evs) \;\triangleq\; \{X\} \cup \mathsf{spies}\,evs$$

2. Any message noted by a compromised agent in a trace is known to the Spy on that trace.

$$\mathsf{spies}\,((\mathsf{Notes}\,A\,X)\,\#\,evs) \;\triangleq\; \begin{cases} \{X\} \cup \mathsf{spies}\,evs & \text{if } A \in \mathsf{bad} \\ \mathsf{spies}\,evs & \text{otherwise} \end{cases}$$

It can be seen that $\mathsf{spies}\,evs$ contains the entire network traffic that occurred during the history recorded by evs (it actually is a larger set as it contains some notes) and therefore will be often addressed as *traffic on evs*. By appropriately referring to this set, we meet requirement 2, that the Spy control the network traffic.

Meeting requirement 3 of the implementation of Dolev-Yao's threat model requires the implementation of suitable message operators, which we prefer to treat separately in the next section.

3.10 Operators

Three operators are introduced to manage sets of messages, all sharing the same declaration

analz, synth parts : msg set $\longrightarrow$ msg set

The first operator formalises the act of breaking up messages into their components. If H is a set of messages, then analz H can be defined inductively as follows.

0. Any element of a given message set can be analysed from it.

   $X \in H \Longrightarrow X \in$ analz H

1. The first component of a concatenated message that can be analysed from a given message set can be itself analysed.

   $\{X, Y\} \in$ analz $H \Longrightarrow X \in$ analz H

2. The second component of a concatenated message that can be analysed from a given message set can be itself analysed.

   $\{X, Y\} \in$ analz $H \Longrightarrow Y \in$ analz H

3. The body of a ciphertext that can be analysed from a given message set can be itself analysed provided that the encrypting key can be analysed too.

   $[\![$ Crypt $KX \in$ analz H; Key(invKey$K$) $\in$ analz H $]\!] \Longrightarrow X \in$ analz H

The set analz(spies evs) is particularly important. It is the set of all message components that the Spy can derive from the observation of the traffic in the trace evs. According to the definition of analz, those components are obtained by decomposing concatenated messages and opening up ciphertexts using keys that become recursively available. But no ciphertext can be opened without the corresponding key. Therefore, we use

$X \notin$ analz(spies evs)

to formally state confidentiality of X in the trace evs.

The operator synth formalises the act of building up new messages from known components. If H is a set of messages, then synth H can be defined as follows.

0. Any element of a given message set can be synthesised from it.

   $X \in H \Longrightarrow X \in$ synth H

1. Any agent name can be synthesised from any message set.

   Agent $A \in$ synth H

2. If a message can be synthesised from a given message set, then so can its hash.

$X \in \mathsf{synth}\, H \implies \mathsf{Hash}\, X \in \mathsf{synth}\, H$

3. If two messages can be synthesised from a given message set, then so can their concatenation.

$[\![X \in \mathsf{synth}\, H;\ Y \in \mathsf{synth}\, H \,]\!] \implies \{\!|X, Y|\!\} \in \mathsf{synth}\, H$

4. If a key belongs to a given message set and a message can be synthesised from it, then so can the encryption of the message with the key.

$[\![\mathsf{Key}\, K \in H;\ X \in \mathsf{synth}\, H \,]\!] \implies \mathsf{Crypt}\, K X \in \mathsf{synth}\, H$

Observe that, while rule 1 makes all agent names trivially available for the Spy to form new messages, no analogous rule does that for nonces. Hence, nonces remain unguessable for the Spy. The set $\mathsf{synth}(\mathsf{analz}(\mathsf{spies}\ evs))$ contains all messages that the Spy can synthesise by concatenation and encryption using components derived from the traffic on evs. It meets requirement 3 of the implementation of Dolev-Yao's threat model (§3.9), that the Spy can perform any message operation except cryptanalysis.

The operator parts extracts all components from a given set of messages by projection and decryption. Given a set H, the set $\mathsf{parts}\, H$ can be defined as follows.

0. Any element of a given message set is a component of the given message set.

$X \in H \implies X \in \mathsf{parts}\, H$

1. The first component of a concatenated message that is a component of a given message set is itself a component of the given message set.

$\{\!|X, Y|\!\} \in \mathsf{parts}\, H \implies X \in \mathsf{parts}\, H$

2. The second component of a concatenated message that is a component of a given message set is itself a component of the given message set.

$\{\!|X, Y|\!\} \in \mathsf{parts}\, H \implies Y \in \mathsf{parts}\, H$

3. The body of a ciphertext that is a component of a given message set is itself a component of the given message set.

$\mathsf{Crypt}\, K X \in \mathsf{parts}\, H \implies X \in \mathsf{parts}\, H$

The definitions of analz and parts are similar. However, contrarily to the former, the latter does not require the encrypting key to extract the body of a ciphertext; hence, parts somewhat formalises unlimited computational power. Thus, given a message X and a trace evs,

$X \in \mathsf{parts}(\mathsf{spies}\ evs)$

signifies that X appears in the traffic on *evs*, possibly as a component of a larger message. A simple fact involving these two operators is

analz $H \subseteq$ parts H

and it can be easily proved. When H is the traffic in a trace, the theorem signifies that the traffic components that are available to the Spy on *evs* form a subset of all traffic components.

Another important concept for security protocols is *freshness*. For example, when an agent has to issue a nonce or a session key, it is good security practice to generate a fresh one, namely one that has never been used before, so that it cannot collide with any other keys. We introduce the function

used : event list $\longrightarrow$ msg set

formalising the set of all message components that either belong to some agent's initial knowledge or appeared on some trace. It can be defined by primitive recursion as follows.

0. All components of any agent's initial state are used on any trace, including the empty one.

used $[] \triangleq \bigcup B.$ parts(initState B)

1. All components of a message sent in a trace are used on that trace.

used$((\text{Says } A\,B\,X) \# evs) \triangleq$ parts$\{X\} \cup$ used *evs*

2. All components of a message noted in a trace are used on that trace.

used$((\text{Notes } A\,X) \# evs) \triangleq$ parts$\{X\} \cup$ used *evs*

Freshness of a message m in a trace *evs* can be formalised by $m \notin$ used *evs*.

3.11 Protocol Model

The formal protocol model is a set of lists of events, namely a set of traces, and is unbounded because so are Isabelle lists. It contains all traces such that each formalises a possible history of the network while the protocol is in use. Since it is defined inductively, we remark that it contains any possible trace induced by the protocol, including the most peculiar ones such as a trace that sees a single agent initiate a large number of protocol sessions but no replies at all.

The base case of the definition states that the empty trace belongs to the set. Then, each inductive rule formalises a protocol step. More precisely, it states how to extend a given trace of the set by the event formalising the protocol step. If a protocol has n steps, its model at least has n inductive rules, each introducing a Says event.

Another rule called *Fake* lets the Spy send all messages that she can fake: if *evs* is a trace of the set, then also the concatenation of Says Spy B X with *evs* must be a trace of the set; here, X is taken from synth(analz(spies *evs*)), a set we have discussed in the previous section. Hence, this rule contributes to the implementation of Dolev-Yao's threat model. When analysing smartcard protocols (Chapter 11) or second-level protocols (Chapter 13), we will see that an extra rule is required to model the Spy's activity. That rule will extend the threat model by allowing the Spy to construct a session key out of known components using a public algorithm. A similar concept can be found in the inductive analysis of TLS [134].

An important extra rule will be used to model security protocols that distribute a session key. Called *Oops*, it models an agent's loss, either deliberate or not, of a session key to the Spy. The rule introduces an event, referred to as an *oops event* below, whereby the Spy notes down the session key by a Notes event. This makes it possible to investigate how local breaches of security can affect the global security. We shall see various forms of such a rule throughout this book.

$$
\begin{array}{llll}
1. & A \longrightarrow B & : & A, Na \\
2. & B \longrightarrow A & : & \{Na, Kab\}_{sK_B^-}
\end{array}
$$

```
Nil:
[] ∈ dsp

Fake:
⟦ evsF ∈ dsp; X ∈ synth(analz(spies evsF)) ⟧
⟹ Says Spy B X # evsF ∈ dsp

DSP1:
⟦ evs1 ∈ dsp; Nonce Na ∉ used evs1 ⟧
⟹ Says A B {Agent A, Nonce Na} # evs1 ∈ dsp

DSP2:
⟦ evs2 ∈ dsp; Key K ∉ used evs2; K ∈ symKeys;
  Says A' B {Agent A, Nonce Na} ∈ set evs2 ⟧
⟹ Says B A (Crypt (priSK B) {Nonce Na, Key K}) # evs2 ∈ dsp

Oops:
⟦ evsO ∈ dsp;
  Says B A (Crypt (priSK B) {Nonce Na, Key K}) ∈ set evsO ⟧
⟹ Notes Spy {Nonce Na, Key K} # evsO ∈ dsp
```

Fig. 3.3. Example protocol and corresponding inductive model

For the sake of demonstration, Figure 3.3 shows a demo security protocol along with its inductive model. The protocol sees an initiator send her identity and a fresh nonce to a responder, who replies by encrypting the nonce and a fresh session key with his private signature key.

The constant dsp denotes the inductive model of the demo security protocol. It is defined by five rules that indicate which traces the protocol can produce. Rule *Nil* sets the base of the induction stating that the empty trace belongs to the protocol model. Rule *Fake*, as explained, allows the Spy to generate fake messages using material gleaned from past traffic, and to send it to anyone.

The postcondition of each inductive rule confirms that traces are built in reverse chronological order because new events are added at the head of the given trace. Also, it can be seen that each rule refers to a generic trace; for example, *Fake* refers to *evsF*, *DSP1* refers to *evs1*, and so on. Because there is no bound between variables of different rules, the trace names could have been identical. However, with long proofs especially, the use of typical trace variable names helps us identify which rule each subgoal arises from.

Rule *DSP1* specifies the first protocol step, whereby A chooses a fresh nonce and sends it to B. The rule follows the usual inductive pattern: given a trace *evs1* of the protocol model, then its extension with the event formalising the first step of the protocol is still a trace of the protocol model. Observe the implementation of freshness of the nonce in terms of the function used. The rule has very few preconditions, which signifies that anyone can initiate a protocol session with anyone else at anytime, provided that she can access the random number generator to issue a fresh nonce.

Rule *DSP2* models the second protocol step, which is inherently conditional: if anyone sent B an instance of the first message, then B replies with the second. The rule follows this conditional structure by having the event modelling the first step of the protocol among its preconditions. Precisely, *DSP2* states that, if any A' sent B a suitable message during the protocol history that *evs2* models, then B signs the nonce just found and a fresh session key and sends the outcome to A. He cannot verify that the message came from A because the originator of a message is never obvious under our threat model, but replies to the agent named by the first component of the received message. We will not formalise message reception until later (§8.2). Observe the implementation of a fresh session key.

Our demo protocol distributes a session key, so we define an *Oops* rule. It states that whenever B issues a session key, with the event formalising the second step of the protocol, the Spy can note it down along with the nonce corresponding to the same session. If we recall the definitions of spies and of analz from the previous section, it is clear that on any trace *evs* in which the *Oops* rule fires, we have that both the session key and the corresponding nonce belong to analz(spies *evs*).

Having seen all protocol rules, it is clear that not every conceivable trace belongs to the protocol model. For example, no trace where an agent other than the Spy notes down a message does, because no rule introduces such an event. In consequence, the example trace seen above in Figure 3.2 does not belong to the inductive model of our example protocol as it stands. If we want to transform it, as an exercise, into an admissible trace, we must certainly prune it of event Notes B (Nonce Nc). Also, no rule allows the Spy to note down a single nonce; hence, event Notes Spy (Nonce Na') must be deleted too. This means that the Spy never learns nonce Na'. Therefore, event Says Spy D (Nonce Na') is not admissible also because the Spy cannot perform it even when acting legally, namely in accordance with rules *DSP1* and *DSP2*. Figure 3.4 presents the result of our exercise: a trace that belongs to the model of our example protocol. It can be seen that each event can be introduced by one of the inductive rules defining the protocol model. For example, the last event is introduced by the *Fake* rule.

Says Spy D ⦃Agent A, Nonce Nc⦄

Notes Spy ⦃Nonce Nc, Key Kcb⦄

Notes Spy ⦃Nonce Nc, Key Kcb⦄

Says B C (Crypt (priSK B) ⦃Nonce Nc, Key Kcb⦄)

Says C B ⦃Agent C, Nonce Nc⦄

Says A B ⦃Agent A, Nonce Na'⦄

Says A B ⦃Agent A, Nonce Na⦄

Fig. 3.4. Example trace belonging to the inductive model of the example protocol

In terms of verification of properties, it is useful that the properties of an inductively defined set can be established by the corresponding induction principle: each putative fact must be verified against all rules defining the set. This generates long case analyses, which Isabelle helps to solve mechanically. A property of the set typically attempts to formalise some protocol goal, as we will discuss extensively in the next chapter.

4. Verifying the Protocol Goals

Several goals of security protocols, which are only part of the overall security of a system, are discussed along with the strategies to prove them using the Inductive Method. However, a few extensions are necessary to deal with additional though fundamental protocol goals.

Security protocols cover a large spectrum of applications, ranging from e-mail to banking services, and are intended to achieve a number of goals that depend on the specific application. Today, we take it for granted that "security is not a simple boolean predicate" [12] although this has not always been obvious. Security can be thought of as a *conjunctive normal form* formula. Only some of its conjuncts represent protocol goals, while the remaining ones embody properties of the entire system that adopts the protocol. Devising the right "security formula" for a specific application is matter of open research. It is not obvious how many and which conjuncts the security formula should have. For example, many LANs have suffered breaches of security due to badly configured firewalls and of some important secrets saved in the clear on internal workstations. Those secrets were appropriately protected by established security protocols whenever they were sent over the network. However, external attackers managed to read the secrets from the workstations by merely trespassing the firewall. This demonstrates that security is a concept that spreads across multiple vertices of a communication architecture and multiple levels of each vertex.

Even verifying whether a single conjunct of a putative security formula holds may be a challenging task. In particular, only at the beginning of the last decade was attention devoted to the conjuncts representing the goals of security protocols through the use of formal methods. For example, an agent may need evidence that a received message is reliable in all its components (*integrity*), or that the same session key cannot be received within two different messages (*unicity*), or that his peer is indeed meaning to communicate with him (*authentication*).

The Inductive Method, introduced in the previous chapter, supports a scrupulous formal account of these properties. This chapter completes the presentation of the method by explaining how it copes with the main security goals. First, it comments on how to verify the reliability of the protocol model by means of suitable theorems (§4.1) and introduces the regularity lemmas (§4.2). Then, it discusses the goals of authenticity, stressing its correlation with integrity (§4.3). It moves on to unicity (§4.4), confidentiality (§4.5), authentication (§4.6) and key distribution (§4.7). The strategies for proving these goals with the help of the theorem prover Isabelle are presented throughout, along with a few limitations of the original Inductive Method that become clear along the way. The treatment here is as general as possible, though it is often necessary to refer to specific protocol theorems. There are also a few references to the concept of an agent's *minimal trust*, which prepares the ground for the principle of *goal availability*; both concepts are treated separately in the next chapter.

4.1 The Reliability of the Protocol Model

The problem of how close a model is to the system it should represent affects any formal verification. If a model is oriented to the formal establishment of guarantees (theorems), then suitable *reliability* theorems can highlight whether it suffers any discrepancies with the real system or, rather, whether it behaves as it is expected to. Although these theorems may not address any protocol goals directly, they can dramatically increase the real-world significance of any other theorems proved subsequently of the model. For example, special reliability theorems establish properties that hold until a certain event takes place in a trace. For this purpose, we shall define the function before in this section.

The theorems that fall under this category cannot be enumerated exhaustively, as new ones may arise from specific protocols to analyse. Many basic results available with theory files Message.thy, Event.thy and Public.thy (Figure 3.1) certainly count as reliability theorems. For example, if the Spy has obtained a set H of messages from the observation of the traffic, she may extract message components from them by decomposing concatenated messages and decrypting the ciphers sealed with known keys. This process can be iterated until there are no more concatenated messages and no more intelligible ciphers. In the real world, at this stage, the Spy cannot acquire new knowledge by repeating the previous process. The model conforms to this, as the analz operator, which performs the complete message analysis, can be proved to be idempotent: $\mathsf{analz}(\mathsf{analz}\,H) = \mathsf{analz}\,H$.

Another important reliability result states that a cryptographic key that is fresh in a trace is certainly not a long-term (shared) key. This assures us that when a fresh key is generated, it cannot clash with any agent's shared key. Indeed, the probability of this happening in the real world is negligible.

A *possibility property* [133, §4.1] states that the protocol model contains a trace in which the event formalising the last step of the real protocol occurs. This form of weak liveness must be considered as a reliability theorem because it signifies that the model allows completion of the protocol.

Other theorems of this class state that if an agent other than the Spy sends a certain message of the protocol, then the components of the message can be specified with certainty. Indeed, these agents act in the real world according to known, predefined rules. For example, the Server of the shared-key Needham-Schroeder protocol (Figure 2.6) can be proved to send only well-formed messages [33, 34]. If *evs* is a trace of the protocol model containing

Says Server A (Crypt $K'\{\!|Na,$ Agent $B,$ Key $Kab,\ ticket|\!\}$)

then

$K' = $ shrK A and $Kab \notin $ range shrK   and
$ticket = $ Crypt(shrK $B)\{\!|$Key $Kab,$ Agent $A|\!\}$

The full proof script consists of three Isabelle commands: the first makes the event of the assumption a premise of the inductive formula, the second applies induction, and the third simplifies all subgoals. It comes with the theory file NS_Shared.thy (Figure 3.1). However, this theorem does not guarantee that the Server's operation is entirely reliable. For example, it is not clear whether the issued session key is fresh as it should be. To investigate this formally, we declare the function

before : [event, event list] $\longrightarrow$ event list

so that before *ev evs* yields the subtrace of events that occur in the trace *evs* before the introduction of the event *ev*. Since all traces are extended in reverse, the head of a trace contains the most recent events. So, the definition scans the trace *evs* reversed through function rev, and collects its elements until *ev* is found through function takeWhile

before *ev* on *evs* $\triangleq$ takeWhile$(\lambda z.z \neq ev)$(rev *evs*)

Functions rev and takeWhile come from the theory file of lists List.thy [33, 34]. On the assumptions of the previous theorem, we have proved that the session key is fresh when the Server issues it, formally that

Key $Kab \notin$ used(before
     (Says Server A (Crypt $K'\{\!|Na,$ Agent $B,$ Key $Kab, X|\!\}$))
     on *evs*)

which completes the argument about the reliability of the model Server. The proof requires two general subsidiary lemmas, one stating that the set of elements used in a trace is the same as the set of elements used in the reversed trace, the other stating that, if an element is used on a subtrace, then it is also used in the trace from which the subtrace is derived.

When verifying protocols based on smartcards (§11.3.1), we will prove that any agent other than the Spy can use only his own smartcard, provided that it has not been stolen by the Spy. By contrast, the Spy can use both her own card and a set of compromised cards. These are also reliability theorems.

Proving a theorem of this class is not difficult. The idempotence of analz is easily derived from its definition, while the possibility property is proved by "joining up the protocol rules in order and showing that all their preconditions can be met" [133, §4.1]. All remaining theorems, strictly depending on the specific protocol being verified, necessitate induction over the protocol rules. Then, the simplifier either terminates all subgoals or, alternatively, highlights the structure of the remaining ones. On these occasions, Isabelle's proof method auto, which combines simplification and classical reasoning, often concludes the proof.

4.2 Regularity

Broadly speaking, *regularity* lemmas are facts that can be proved for a message that appears in the traffic [133, §4.3]. Precisely, if a trace *evs* of the protocol model is such that

$$X \in \mathsf{parts}(\mathsf{spies}\ evs)$$

then some facts can be proved about X and *evs*. Such a broad definition includes many of the theorems proved for a protocol model. In particular, it includes the theorems formalising the goals of authenticity and authentication (§§4.3 and 4.6). We take a more restrictive view, and address as regularity lemmas only the *key regularity* ones.

A key regularity lemma holds for a long-term key that a protocol never sends in the traffic. Formally, given a trace *evs* of the protocol model, an agent's key appearing in the traffic on *evs* implies that the agent is compromised. The implication considers the fact that it must have been the Spy who sent the key in the traffic because the owner is assumed to never do so. The proof applies induction and shows that the key could appear in the traffic only in the *Fake* case, namely when it is the Spy who uses it, because the key owner is compromised. By definition of initState, spies and parts, the same result holds in the opposite direction regardless of the protocol being analysed. Combining the two implications, the key regularity lemma for shared keys reads as

$$\mathsf{Key}(\mathsf{shrK}\ A) \in \mathsf{parts}(\mathsf{spies}\ evs)\ \text{ if and only if }\ A \in \mathsf{bad}$$

The same lemma is even more relevant if expressed in terms of analz as

$$\mathsf{Key}(\mathsf{shrK}\ A) \in \mathsf{analz}(\mathsf{spies}\ evs)\ \text{ if and only if }\ A \in \mathsf{bad}$$

The left-to-right implication holds because of $\mathsf{analz}\ H \subseteq \mathsf{parts}\ H$ and the key regularity lemma; the opposite implication holds by definition of initState, spies and analz. Its importance lies in translating a condition that an agent is certainly not able to verify — the Spy's learning his key from the analysis of the traffic — into one that he can verify (Chapter 5) — his being compromised. In other words, the lemma says that the Spy learns a shared key from analysing the traffic induced by the protocol if and only if she knows it initially. For example, let us consider a certificate meant for A and sealed with her shared key. If we want to prove any properties about the certificate, then it must be tamperproof against the Spy. This in turn requires A's key not to be available to the Spy, which, by the regularity lemma, is equivalent to A's being uncompromised.

We remark that regularity lemmas can be proved for any long-term keys or secret nonces that are never sent in the traffic. For example, a suitable form holds for smartcard keys with most smartcard protocols (§11.3.2), and of private signature or encryption keys with most accountability protocols (§13.2.2).

4.3 Authenticity

If a message that appears to have originated with a certain agent did indeed originate with that agent, then the message enjoys *authenticity*. The ISO Security Architecture framework [93] distinguishes authenticity from *integrity*, which holds for a message that is proved to be received in the same form as the one in which it was generated.

However, many researchers consider the source of a message as an essential part of the message. Therefore, verifying that the message is unaltered when it is received (integrity) confirms its originator (authenticity). Conversely, verifying the originator of a message that is received also confirms that the message is unaltered. To this extent, the two properties may be considered equivalent.[1]

Our proofs support this viewpoint. For example, let us consider the ticket $\{Kab, A\}_{Kb}$ of the shared-key Needham-Schroeder protocol (Figure 2.6), which is created by the Server. The integrity of the ticket is equivalent to preventing the Spy from learning Kb and thus, via the regularity lemma, to B's being uncompromised. Under this assumption, the ticket can be proved to have originated with the Server, and hence it is authentic [33, 34]. So, the ticket integrity implies its authenticity.

The converse also holds: if the ticket integrity fails, then so does the ticket authenticity. Let us assume that integrity fails. It follows that the Spy knows Kb and then, by the regularity lemma, that B is compromised. Trying to prove authenticity in this scenario leaves the subgoal in Figure 4.1, which arises from the case *Fake*.

The second and third assumptions signify that, although the ticket does not appear in the traffic, the Spy can synthesise it from the analysis of that traffic. The symbolic evaluation of synth at this stage says that two cases

---

[1] Private conversation with Dieter Gollmann.

⟦ evsF ∈ ns_shared; B ∈ bad;
  Crypt (shrK B) ⦃Key Kab, Agent A⦄ ∉ parts (spies evsF);
  Crypt (shrK B) ⦃Key Kab, Agent A⦄ ∈ synth (analz (spies evsF)) ⟧
⟹ ∃ Na. Says Server A
              (Crypt (shrK A) ⦃Na, Agent B, Key Kab,
                                Crypt (shrK B) ⦃Key Kab, Agent A⦄⦄)
          ∈ set evsF

Fig. 4.1. Proving ticket authenticity without ticket integrity for shared-key
Needham-Schroeder: failed

are possible. Either the Spy merely forward the ticket that she obtains from
the analysis of the traffic, namely the ticket belongs to analz(spies *evsF*); or
the Spy handle all components necessary to forge the ticket. The former case
is impossible because it contradicts the second assumption of the subgoal,
following analz $H \subseteq$ parts H. According to the latter case, simplification and
the regularity lemma transform the third assumption in the pair

Key $Kab \in$ analz(spies *evsF*) and $B \in$ bad

The resulting subgoal can be falsified because it assumes a non-contradict-
ory scenario in which an agent and a session key are compromised to the Spy.
The proof of ticket authenticity fails: the Spy can forge the ticket before the
Server issues it legally.

In light of these considerations, we will regard authenticity and integrity
as a single concept denoted by the former term. Also, when proving message
authenticity, we will in general make the assumptions that appear to prevent
the Spy from faking the message, and will attempt to enforce the event corresponding to the protocol step that creates the message. If we are dealing
with a certificate sealed with a long-term key, then, by application of the
corresponding regularity lemma, it will suffice to assume that the key owner
is uncompromised. However, further assumptions may be required to investigate the authenticity of specific message components such as the "pairkey"
used by the Shoup-Rubin protocol (§11.3.3).

The introduction of message reception (§8.2) will allow for a more realistic
formalisation of the authenticity theorems from the agents' viewpoints, as
prescribed by the principle of *goal availability*, introduced in the next chapter.

4.4 Unicity

Security protocols often involve the creation of fresh components such as
nonces or session keys. A fresh nonce can be used to uniquely identify a
protocol session, while a fresh session key shall not be used beyond the current
session.

It follows that freshness is somewhat a synonym with unicity. Precisely,
a fresh message component is uniquely bound to its message of origin. This

observation inspired the *unicity* theorems [133, §4.4]. In its original formulation, a typical unicity theorem establishes that, if two events containing a fresh message component occur, then the events are identical. Various theorems of this form will be presented throughout this book, pertaining to either nonces or session keys. We also advance a novel formulation indicating that the event containing a fresh message component cannot occur more than once. It relies on the predicate Unique, introduced below.

To begin with an example of the original formulation, let us consider the Yahalom protocol (Figure 8.8) It requires the Server to issue a fresh session key Kab for two peers A and B [135]. So, if evs is a trace of the Yahalom model containing the events

Says Server A {Crypt(shrK A){Agent B, Key Kab, Na, Nb}, X} and

Says Server A' {Crypt(shrK A'){Agent B', Key Kab, Na', Nb'}, X'}

then

$A = A'$ and $B = B'$ and $Na = Na'$ and $Nb = Nb'$

Another example of the original formulation of a unicity theorem can be derived from the public-key Needham-Schroeder protocol (Figure 2.5). An initiator A of this protocol has to issue a fresh nonce Na and include it in the first message. Therefore, if that nonce is not available to the Spy and evs is a trace of the protocol model such that

Crypt(pubK B){Nonce Na, Agent A} ∈ parts(spies evs) and

Crypt(pubK B'){Nonce Na, Agent A'} ∈ parts(spies evs)

then

$A = A'$ and $B = B'$

Conversely, should Na be available to the Spy, she might have created at will the certificates containing the nonce, even for random agents A and A', and the conclusion would not hold. So, an alternative theorem relies on both B and B' being uncompromised. Proving these theorems requires an inductive analysis of the protocol steps and a simplification of the arising subgoals. The critical case is the step where the fresh component is created. If a different component is introduced, then simplification terminates the subgoal; if the same component is introduced, then the freshness assumption concludes the proof.

The first unicity theorem discussed above is only useful to the Server, who is the only entity capable of checking its assumptions — no other agent can directly verify that the Server issues certain messages. The second theorem can be applied by an uncompromised agent B only upon reception of the certificates {Na, A}$_{Kb}$ and {Na, A'}$_{Kb}$. Once B has received the two certificates, should A differ from A', the theorem will be violated. Thus, B will suspect that something that lies outside the formal model has happened, ranging from some failure of the underlying transport protocol to brute-force codebreaking. However, these considerations mentioning reception are informal. For the theorem to attain a higher level of formal expressiveness, we shall introduce message reception in the model (§8.2).

Analysing the design of Kerberos V (Chapter 9) leads us to study the relation between timestamps and unicity for the first time. We realise that uncompromised agents always pick the right timestamp, which is the current time. Because our formalisation of timestamping (Chapter 6) prevents two events from occurring at the same time, it follows that an uncompromised agent always inserts fresh timestamps in the messages he creates. This implies, by the same reasoning conducted over nonces and session keys, that the corresponding events are unique.

We investigated the unicity argument further. Its original formulation allows a fresh component to appear more than once within identical messages. For example, the unicity theorem for the Yahalom Server would not be violated if the Server sent a fresh session key within two equal messages. We realise that this case can be ruled out thanks exactly to the freshness assumption. However, it must be expressed formally. We declare a predicate that takes as parameter an event and a trace. It is

$$\mathsf{Unique} : [\mathsf{event}, \mathsf{event\ list}]$$

It scans the trace until the event is found and skipped, and then checks that the event does not occur on the rest of the trace

$$\mathsf{Unique}\ ev\ \mathsf{on}\ evs\ \triangleq\ ev \notin \mathsf{set}(\mathsf{tl}(\mathsf{dropWhile}(\lambda z.z \neq ev)\ evs))$$

The predicate holds on an event ev and a trace evs when ev occurs only once on evs. Precisely, function dropWhile scans the trace until the event is found, and then function tl prunes it — these two functions come from the theory file List.thy [33, 34]. As expected, we can prove that a trace evs of the Yahalom model is such that

$$\mathsf{Unique}\ (\mathsf{Says\ Server}\ A\ \{\!|\mathsf{Crypt}(\mathsf{shrK}\ A)\{\!|\mathsf{Agent}\ B, \mathsf{Key}\ Kab, Na, Nb|\!\}, X|\!\})$$
$$\mathsf{on}\ evs$$

An equivalent result can be routinely proved by induction and simplification for all protocols analysed so far. The definition of the predicate must be used as a rewrite rule for the simplifier. In particular, the analysis of protocols that are based on smartcards (§11.3.4) will gain by the new theorem in the case where the protocols assume a secure means between agents and cards. Agents will receive additional guarantees on the functioning of the cards.

4.5 Confidentiality

In the Dolev-Yao threat model, a protocol enforces *confidentiality* of a message m if it does not disclose m to the Spy.

Let us focus on session key confidentiality on the original Otway-Rees protocol (Figure 8.4). If A and B are uncompromised and *evs* belongs to the Otway-Rees model and contains

Says Server B {|Na, Crypt(shrK A){|Na, Key Kab|},
    Crypt(shrK B){|Nb, Key Kab|}|}

but does not contain an oops event on Kab involving the same nonces

Notes Spy {|Na, Nb, Key Kab|}

then *evs* is such that

Key $Kab \notin$ analz(spies *evs*)

Proving this Theorem [133, §4.6] requires evaluating the assertion for all the possible forms of *evs* according to the protocol model. In each of these cases, spies extracts the new message, say X, leaving expressions of the form

Key $Kab \notin$ analz({X} $\cup$ (spies *evs*))

The symbolic evaluation rules for analz inspect X and pull out all components except the keys. For example, they leave an expression of the form

Key $Kab \notin$ analz({Key K} $\cup$ (spies *evs*))

in the subgoal corresponding to the oops event. Here, K is the non-fresh session key that the oops event introduces. Symbolic evaluation cannot proceed any further because in general K might be used to encrypt Kab in some message appearing on *evs*. However, security protocols usually prevent this, and Otway-Rees makes no exception. Proving such a result, addressed as the *session key compromise theorem* [133, §4.5], provides the necessary rewriting rule

Key $K' \in$ analz({Key K} $\cup$ (spies *evs*)) if and only if
$K' = K$ or Key $K' \in$ analz(spies *evs*)

where K is a session key. The result confirms that Otway-Rees never uses session keys to encrypt other keys, so the Spy cannot exploit a stolen session key to learn others. Technically, it is easier to prove the theorem first for a generic set of session keys, and then to specify it for a single session key.

After simplification by the compromise theorem, the oops subgoal of the secrecy theorem terminates via an appeal to the unicity argument: the Server issues Kab only with the nonces Na and Nb. The subgoal corresponding to the third step of the protocol, where the Server issues the session key, is solved by freshness of the session key, and the remaining subgoals routinely.

Although the session key confidentiality theorem constitutes the main confidentiality result, it is still not directly applicable by the agents. Further lemmas are necessary for this purpose (§2.2.4).

Confidentiality is often crucial also on nonces. For example, in the TLS protocol [134], the pre-master secret, *PMS*, is a nonce of fundamental importance because it is used to compute other nonces, session keys and MACs. Confidentiality of the *PMS* can be proved conventionally with the addition of a new rewriting rule for the analz operator. If A and B are uncompromised and *evs* is a trace of the TLS model containing

Notes A {|Agent B, Nonce *PMS*|}

then *evs* is such that

Nonce $PMS \notin$ analz(spies *evs*)

The theorem signifies that once an uncompromised agent notes the *PMS*, it remains secure from the Spy. As explained on the preceding result, the proof requires a suitable rewriting rule for analz, stating that session keys cannot be exploited to learn nonces. If K is a session key, then *evs* is such that

Nonce $N \in$ analz({Key K} $\cup$ (spies *evs*)) if and only if
Nonce $N \in$ analz(spies *evs*)

As we shall see in the following, the methods presented in this section remain unvaried after the introduction of message reception or of smartcards into the formal treatment.

4.6 Authentication

Despite the fact that agent *authentication* is the main, claimed goal of many security protocols, there exists significant potential for confusion about the interpretation of this term [81]. A taxonomy due to Lowe may elucidate the matter identifying four levels of authentication. Let us suppose that an initiator A completes a protocol session with a responder B.

1. *Aliveness of B* signifies that B has been running the protocol.
2. *Weak agreement of B with A* signifies that B has been running the protocol with A.
3. *Non-injective agreement of B with A on H* signifies weak agreement of B with A and that the two agents agreed on the set H of message components.
4. *Injective agreement of B with A on H* signifies non-injective agreement of B with A on H and that B did not respond more than once on each session with A.

Observe that each level subsumes the previous one. In particular, the injective agreement of B with A establishes an injective relation between B's runs of the protocol with A and A's runs with B. The existing authentication arguments carried out using the Inductive Method do not set their findings within this taxonomy. Although authentication is typically interpreted as aliveness, we find that many guarantees also convey weak agreement. However, investigating non-injective agreement requires extending the approach, as we explain below (Chapter 8).

The Inductive Method allows agents to respond more than once to a received message (provided that it is in the expected form) because the protocol models are meant to be as permissive as possible. Therefore, the strongest form of authentication that we can wish to prove is non-injective agreement. A generic rule of the protocol model takes a trace evs of the model, insists that some events occur on it and others do not, and establishes that the concatenation of a specific event ev with evs is still a trace of the model. Constraining agents to a single reply can be done by adding to the generic rule the extra assumption that ev does not occur on evs. While leaving the existing proofs unaltered, this would make it possible to investigate injective agreement. Lowe simply says that such a property "may be important in, for example, financial protocols" [109, §2.4], but we are not aware of real-world protocols that have claimed it explicitly as a goal. So, it does not raise particular interest from the verification standpoint.

Most importantly, we realise that the original Inductive Method requires some extensions to even formalise non-injective agreement. With key distribution protocols, for example, non-injective agreement on a session key is the relevant form of authentication. Let us consider the original guarantees proved about the shared-key Needham-Schroeder protocol (Figure 2.6), which came with Isabelle's earlier distributions up to Isabelle98-1 [158]. If A and B are uncompromised and evs, which is a trace of the formal protocol model, contains

Says $B\ A\ (\mathsf{Crypt}\ Kab(\mathsf{Nonce}\ Nb))$

and is such that

$\mathsf{Crypt}(\mathsf{shrK}\ B)\{\!|\mathsf{Key}\ Kab, \mathsf{Agent}\ A|\!\} \in \mathsf{parts}(\mathsf{spies}\ evs)$ and
$\mathsf{Crypt}\ Kab\{\!|\mathsf{Nonce}\ Nb, \mathsf{Nonce}\ Nb|\!\} \in \mathsf{parts}(\mathsf{spies}\ evs)$

but does not contain an oops event on Kab, for any N and N', of the form

Notes Spy $\{\!|N, N', \mathsf{Key}\ Kab|\!\}$

then evs contains

Says $A\ B\ (\mathsf{Crypt}\ Kab\{\!|\mathsf{Nonce}\ Nb, \mathsf{Nonce}\ Nb|\!\})$.

The message $\{\!|Nb-1|\!\}_{Kab}$ is formalised as $\mathsf{Crypt}\ Kab\{\!|\mathsf{Nonce}\ Nb, \mathsf{Nonce}\ Nb|\!\}$ (if the model Spy could add or subtract 1, then she could spoof all nonces), so the assertion means that the last step of the protocol has taken place. Reviewing the theorem, we have discovered that the first assumption is superfluous. The general method for all theorems of this form appeals to authenticity (§4.3) and then derives the confidentiality of the session key (§4.5). Hence, if the certificate $\mathsf{Crypt}\ Kab\{\!|\mathsf{Nonce}\ Nb, \mathsf{Nonce}\ Nb|\!\}$ appears in the traffic, then it is integral; so induction proves it to have originated with A. This proof has not required assuming that B sent the other certificate, which can in fact be proved as a corollary.

Does this theorem establish non-injective agreement of A with B on Kab? Upon reception of the two certificates mentioned, and with the assumptions that A is uncompromised and that Kab has not been leaked by accident (which belong to B's *minimal trust*, see Chapter 5), B is informed that A sent him a certificate sealed with Kab. While B learns Kab from one of the certificates, A's merely sending the certificate does not express A's knowledge of the key that seals it. So, B is not informed whether A agrees on Kab. In general, A might be just forwarding an unintelligible message previously received. By Lowe's definitions, the theorem establishes only weak agreement of A with B. However, the informal inspection of the protocol highlights that A is in fact the true creator of the certificate and therefore knows the key to seal it. We will prove such insight formally in Chapter 8 by means of two different strategies.

4.7 Key Distribution

Key distribution is an important goal of many protocols. It is met between two peers who complete a protocol receiving the same session key. Studying it in conjunction with our principle of *goal availability* (next chapter) provides a stronger version of this goal. It requires the peers to gain evidence that they share a session key with each other. Precisely, we have *key distribution to B* when there exists evidence that B learnt the session key. Below, we will adopt this stronger version, which may be known to some researchers as *key confirmation*.

Bellare and Rogaway appear to adopt the weaker definition of key distribution. They state that key distribution is "very different from" agent authentication. "The reason is that entity authentication is rarely useful in the absence of an associated key distribution, while key distribution, all by itself, is not only useful, but it is not appreciably more so when an entity authentication occurs along side. Most of the time the entity authentication is irrelevant" [48, §1.6].

Key distribution and agent authentication are certain to be strictly related. Mutual non-injective agreement on a session key is certain to imply key distribution. Moreover, our proofs always establish key distribution via the authentication argument. This appears to be the only successful method to prove key distribution on all protocols analysed so far. Hence, our findings support the claim that the two goals are equivalent.

To make an example of the last implication, let us recall the second and third steps of the shared-key Needham-Schroeder protocol (Figure 4.2). Upon B's reception of the certificate $\{Kab, A\}_{Kb}$, the only method to show B that A also knows Kab is to derive that the certificate originated with A upon

$$2. \quad S \quad \rightarrow \quad A \quad : \quad \{Na, B, Kab, \{Kab, A\}_{Kb}\}_{Ka}$$
$$3. \quad A \quad \rightarrow \quad B \quad : \quad \{Kab, A\}_{Kb}$$

Fig. 4.2. Shared-key Needham-Schroeder protocol: fragment

reception of the second message of the protocol, which delivered Kab to A. This also establishes non-injective agreement of A with B on the session key. Formalising message reception (§8.2) is, once more, required to express this reasoning formally.

5. The Principle of Goal Availability

A principle of realistic protocol analysis is developed. It complements the principles of prudent protocol design, ultimately contributing to strengthening the protocols. It prescribes that the protocol guarantees be based on assumptions that the protocol participants can verify.

A popular approach to protocol analysis relies on a number of general principles of prudent protocol design [7, 14]. The analysis, which is typically carried out by informal means, studies conformity of the protocols to the principles. Although the principles are claimed to be neither necessary nor sufficient for protocol correctness, this approach has unveiled a number of protocol subtleties.

Some design principles concern, for example, the appropriate use of encryption, and point out that "extra encryption" is not necessarily the same as "extra security." Others pertain to timeliness, pointing out how to use a nonce for freshness purposes. But the principles that have been most often appealed to are perhaps those about *explicitness*. They prescribe that each message be explicit about its contents; otherwise, the agents would be forced to heuristic interpretations that typically become points of convergence of attackers' efforts. Syverson takes a cautionary look at these principles [152], but explicitness still appears to be a significant prerequisite to prudent design.

Our contribution is the development of a principle for the realistic formal analysis of security protocols. Some of the preceding discussion (§2.2.4) provides a useful introduction here. Called *goal availability*, our principle offers a valuable complement to the existing design principles for the common aim of strengthening protocols. It prescribes the development of formal guarantees that a protocol meets its goals and, most importantly, prescribes that those guarantees be used by the protocol participants in practice. While goal availability is meant to directly guide protocol analysis, it may have indirect consequences on protocol design through the findings of the analysis. The experiments conducted thus far appear to confirm that adherence to goal availability is necessary for a formal analysis to be realistically significant. This chapter defines goal availability precisely. In consequence, the informal claim that a protocol meets its goals has a precise, formal meaning in this book: the protocol makes its goals available. Conversely, whenever we formally conclude that a protocol fails to make a goal available, we informally say that the protocol does not meet that goal. A sketch of this principle is already published [27], but its application context is finally clarified and exemplified in this book.

Goal availability says that we should look for formal guarantees based on assumptions that the protocol participants can verify; otherwise, any conclusions would be of no use in the real world because the agents could just not appeal to them. However, our confidentiality argument (§4.5) showed that both peers must be uncompromised in order to prevent the Spy from knowing their shared keys and then decrypting the ciphers that deliver the session key. Likewise, the Spy must not have learnt the key because of the agents' incaution (formalised via an oops event). These assumptions form what we call the *minimal trust* because they cannot be verified by any honest agent. Reinterpreting this concept from a specific agent's viewpoint, we observe that an agent's minimal trust includes assumptions about his peer's behaviour, such as that the peer is uncompromised and has not leaked the key.

Our analysis principle is defined here, while its relation with the design principles is left to subsequent chapters. We argue that checking protocol conformity to goal availability would have anticipated the discovery of the proverbial lack of explicitness in the public-key Needham-Schroeder protocol. A similar finding arises from the Shoup-Rubin smartcard protocol (Chapter 11), thus supporting the claim that checking a protocol for goal availability helps us unveil potential lack of explicitness. However, our principle has a broader impact. For example, it unveils new protocol insights on Kerberos IV (Chapter 7) and V (Chapter 9) and on the Otway-Rees protocol (§8.5.1).

The treatment begins by supporting the necessity to define a threat model before any security claim is made (§5.1). Then, the principle of goal availability is formalised (§5.2) and its past incarnations discussed (§5.3). Finally, its applications are anticipated (§5.4).

5.1 The Need for a Threat Model

Our principle of goal availability must be studied in a specific threat model. In fact, it seems fair to assert that some specification of the attacker's potential must be considered to give sense to any security statement. The mentioned principles of prudent protocol design make no exception: they ought to be studied in a specific threat model.

As seen above (§3.9), the most widely accepted threat model to analyse security protocols is due to Dolev and Yao [75]: the Spy has complete control of the network but can do no cryptanalysis. Therefore, it is sensible to study adherence to the design principles exactly in this threat model, although related variants might also be interesting to examine.

For example, let us consider the popular public-key Needham-Schroeder protocol, which we have seen above (Figure 2.5). It mutually authenticates its participants while delivering secret nonces to them. Establishing whether it conforms to the explicitness principle is difficult. Lowe's middle-person attack on this protocol confirms that the second protocol message lacks explicitness because it fails to mention the identity of the sender.

However, it must be observed that Lowe's answer is significant only if the protocol is analysed in a threat model that is at least as pessimistic as Dolev-Yao's. By contrast, it is interesting to observe that the original protocol does not lack explicitness and hence does not suffer the attack in other threat models. The most significant one probably is a model where the Spy is an outsider: she is not a registered agent and hence cannot initiate the protocol.

Any other less powerful variants of the Dolev-Yao Spy, such as passive eavesdroppers, would let us conclude that the original protocol is not flawed. To just advance a more articulated example, suppose that each pair of agents must pre-execute another strong authentication protocol before they engage in a session of the protocol. This means that each message of the Needham-Schroeder protocol would travel on a pre-authenticated channel, while the peers aim at sharing a session key made of the concatenation of the two exchanged nonces. Lowe's attack clearly would not go through in this case, so that lack of explicitness could not be denounced. Many other, more or less realistic threat models can be conceived to allow some correspondingly realistic claim of protocol security.

In consequence, establishing whether or not a protocol holds to the explicitness principle is subordinated to the given threat model. The same certainly applies to the principle of goal availability. Additional discussions appear in the following chapters each time a protocol is studied in relation to goal availability.

5.2 Goal Availability

We first introduce an abstract version of the principle.

Principle 5.2.1 (Goal Availability — abstract version). *A security protocol ought to make its goals available in practice.*

Informally speaking, a goal is *available in practice* to a protocol peer if there exists some point in the protocol execution when the peer becomes entitled to enjoy the goal. We will be more precise below. It is a formal guarantee that states that the peer gets the goal at that point, so a formal model of the protocol is necessary. What is crucial is that the guarantee must rely on assumptions that the peer is able to verify; otherwise, he cannot apply it. If there exists no such guarantee, we can conclude that the protocol fails to make the goal available to that peer; in other words, the protocol fails to conform to the principle of goal availability for the given goal-agent pair. However, other guarantees may exist, letting us conclude, for example, that the same goal is available to another agent.

$$1. \quad A \;\; \rightarrow \;\; S \;\; : \;\; A, B$$
$$2. \quad S \;\; \rightarrow \;\; A \;\; : \;\; \{\!|A, B, Kab|\!\}_{Ka}$$
$$3. \quad S \;\; \rightarrow \;\; B \;\; : \;\; \{\!|A, B, Kab|\!\}_{Kb}$$

Fig. 5.1. Another example protocol

An example usually clarifies an informal presentation. Let us consider the example protocol in Figure 5.1. It aims at the distribution of confidential session keys to the peers with the help of a trusted Server. Each agent owns a long-term symmetric key, and shares it with the Server alone. Suppose that agent A is not the Spy. Upon A's request, the Server issues a session key Kab and sends it off to both peers within encrypted certificates. This protocol can be analysed using the Inductive Method. A typical confidentiality guarantee may have the following form: the event

Says Server A (Crypt(shrK A){|Agent A, Agent B, Key Kab|})

implies that Kab is confidential. Suppose we could convince ourselves that there exists no other confidentiality guarantee; then we could conclude that the protocol fails to make the goal of session key confidentiality available to peer A. Never during the protocol execution can A verify what is happening at other parts of the network. In particular, A cannot witness that the Server sent him the message containing the session key. In other words, the guarantee is not applicable by A.

However, this protocol does make session key confidentiality available to A because it can be proved that A's reception of message

Crypt(shrK A){|Agent A, Agent B, Key Kab|}

implies that Kab is confidential (because A's shared key is known only to A and to the Server). This guarantee says that A is entitled to consider the session key confidential as soon as she receives a specific message containing that key. By contrast, the first guarantee has only theoretical importance when considered from A's viewpoint. In brief, formal guarantees must be developed from each peer's viewpoint.

As discussed in the previous section, establishing conformity to goal availability demands a threat model. As we have now provided the necessary intuition, we can refine principle 5.2.1 as follows.

Principle 5.2.2 (Goal Availability — refined version). *Given a security protocol, there ought to exist a formal guarantee that the protocol goals are available to the peers in a realistic threat model.*

It is clear that the principle intends to guide the formal analysis of protocols. It is strongly dependent on the chosen threat model, so that some goal might be available within some model though non-available within a more realistic model. It is now necessary to precisely define the concept of *available goal*. Recall that a formal protocol model includes a threat model.

Definition 5.2.1 (Available Goal). *Let $\mathcal{P}$ be a security protocol, P be a formal model for $\mathcal{P}$, g be a goal for $\mathcal{P}$, and A be an agent's name. We say that g is available to A in P if there exists a formal guarantee in P that confirms g and that is applicable by A in P.*

The definition signifies that a goal is available to a peer if there exists an *applicable guarantee* for the peer about that goal. We must define the latter concept precisely.

Definition 5.2.2 (Applicable Guarantee). *Let $\mathcal{P}$ be a security protocol, P be a formal model for $\mathcal{P}$, and A be an agent's name. We say that a formal guarantee in P is applicable by A in P if it is established on the basis of assumptions that A is able to verify in P within her minimal trust.*

The discussion above on the toy protocol is useful here. Definition 5.2.2 says that a guarantee relying, for example, on some agent A's sending or receiving a session key should not be considered applicable by an honest agent B in a model, such as Dolev-Yao's, where each honest agent only sees the messages that he alone sends or receives.

But realistic models also formalise a number of facts that pertain to the environment in which the protocols are executed, and that agents can never verify. For example, no agent can verify that his own and his peer's workstations are secure from trojan horse attacks that would disclose all their secrets. Likewise, no agent can make sure that his remote peer keeps from colluding with intruders, or that a certain point-to-point connection is not broken forever. Therefore, if formalised in the threat model, these facts form what we call the agent's *minimal trust*, something that an honest agent can only take for granted. Definition 5.2.2 tolerates the minimal trust within an applicable guarantee.

Definition 5.2.3 (Minimal Trust). *Let $\mathcal{P}$ be a security protocol, P be a formal model for $\mathcal{P}$, and A be an agent's name. The minimal trust of A is the set of environmental facts formalised in P whose truth values A needs to know but can never verify in practice.*

Precisely, it is clear that no honest agent has complete control over the environment except for his own incoming/outgoing traffic. He can only make assumptions about the rest, which indeed are trust assumptions. It follows that the minimal trust of an agent is the set of environmental facts whose truth value the agent needs to establish in order to enjoy formal protocol guarantees, although in practice he cannot get evidence about that value.

Definition 5.2.3 also has an important consequence. The assumptions of a formal guarantee that we require an agent to be able to verify for the guarantee to be applicable remain only those concerning the protocol messages. In fact, the form of the protocol messages fundamentally influences the strengths and weaknesses of the entire protocol.

In this paper, we adopt a threat model based on Dolev-Yao's where in addition agents' collusion with the Spy is explicitly formalised. Hence, in our model, an agent's minimal trust comprises that his peer is neither the Spy nor an accomplice of the Spy's, and that encryption is perfect. The minimal trust may be extended in case specific protocols, such as those based on smartcards, require extensions to the threat model itself. In general, a large minimal trust may characterise a detailed model, while less realistic formal models, such as one that precludes collusion with the Spy, often take the minimal trust for granted and omit it.

Let us summarise. Goal availability tells us that the formal guarantees must be studied from the agents' viewpoints to check if the peers can apply them within their minimal trust. The protocol analyser routinely derives real-world lessons from idealised analyses. We shall see that studying protocols with respect to goal availability obtains for us novel insights.

5.3 Past Incarnations of Goal Availability

Although goal availability has never been formalised into a principle thus far, it is certain to have influenced some previous works, perhaps implicitly. Here, we advance some comments that are necessarily incomplete due to the large number of formal techniques developed in the last decade. As a general remark, the fact that a protocol is specified from each agent's standpoint does not necessarily imply that any guarantees about that protocol model comply with goal availability. Compliance with goal availability entirely depends on the very statement of the guarantee, which can be influenced by the proof technique in use. Therefore, the analyser must continuously bear in mind this principle while his work unfolds.

For example, an inductive protocol specification of is a set of rules, each expressing a fragment of a specific agent's operation. However, Paulson's initial proof experiments by induction and theorem proving had to primarily focus on developing proof strategies for the confidentiality goal, leaving less consideration for the theorem assumptions [132]. Some of the early proof releases unsurprisingly featured assumptions about the responder that were hardly justifiable from the initiator's viewpoint [158]. Each honest agent's view of the network traffic clearly is limited to what he alone sends or receives. Only later, in studying the authenticity of certificates, did he point out that "agents need guarantees (subject to conditions they can check) confirming that their certificates are authentic" [133], which exactly sounds as a call for goal availability. After spelling out the new principle fully, we checked all existing inductive statements against it and therefore had to upgrade a number of statements and corresponding proofs. It is fair to state that the current proof release [156] only features guarantees that are compliant with goal availability. The upgrade process brought forward some interesting insights that are discussed in the rest of this book.

Burrows et al.'s compact analysis by BAN logic [58] indeed is conducted from each agent's viewpoint and accurately weighs up the preconditions of each formal statement. This is exactly in the spirit of goal availability. For example, their work points out what seem to be inexplicable assumptions for certain proofs to proceed: the analysis of the symmetric-cryptography Needham-Schroeder protocol [58] reveals that it is "unusual" and "dubious" that the responder B can merely assume that the received session key is fresh. The same does not apply to the protocol initiator. Because B cannot verify that assumption, we conclude that the protocol fails to make the goal of freshness of the session key available to B. In consequence, it is well known that an intruder can mount a replay attack on the session key against B.

The importance of the agents' viewpoints appears to be implicit in the analyses by CSP and model checking. For example, Lowe's influential contribution was initially oriented to only finding attacks [107, 112]. Only later was the method tailored to proving protocol goals, and it was natural to define the *process* formalising each agent's sending *signals* solely on the *channels* that the agent can access. This process-centric style of specifying the protocols implicitly shows sensitivity to goal availability. But, as remarked above, our principle is determinant during the actual verification process. It seems fair to state that the proof methods that were subsequently adopted, such as equivalence checking and trace equivalence, inherently comply with goal availability [142]. However, a deeper scrutiny of all published proofs, which is beyond our current goals, may turn out to be interesting.

The principle of goal availability appears to be naturally embedded in Fabrega et al.'s analyses by strand spaces [79]. That method lets the analyser study protocol properties over a *bundle*, which is exactly a collection of actions performed by a single peer. Also recent advances [9] obtained using the provable security approach of Bellare and Rogaway [48] seem to proceed in the direction of goal availaibility: agents are divided into clients and servers and security is defined with respect to each. Approaches whose proofs of correctness are based on equivalence checking, such as the Spi-calculus [5], implicitly embed each agent's viewpoint in purely declarative property specifications. However, adherence to goal availability must be explicitly studied to gain unknown protocol insights, as we shall see in the following chapters.

5.4 Anticipating the Applications of Goal Availability

This section presents some preliminary reasoning in terms of goal availability on the public-key Needham-Schroeder protocol, which was represented in Figure 2.5. More significant examples will follow throughout the book.

Both the original protocol version, subject to Lowe's attack, and the version fixed by Lowe were studied inductively by Paulson in our threat model [33, 34]. Let ns_public denote the model for the flawed, original protocol. It comes with the file NS_Public_Bad.thy [33, 34]. Let *evs* be a generic trace of ns_public. We study whether the goal of Nb confidentiality is made available to B. The strongest relevant guarantee that can be proved is Theorem 5.4.1. The theorem names follow our naming conventions (§1.3.2).

Theorem 5.4.1 (NSP_Spy_not_see_NB). *If A and B are uncompromised and evs contains*

Says B A (Crypt(pubK A){|Nonce Na, Nonce Nb|})

but does not contain, for any C,

Says A C (Crypt(pubK C)(Nonce Nb))

then evs is such that Nonce Nb $\notin$ analz(spies evs).

While honesty of the peers belongs to B's minimal trust, A's refraining from sending any instance of the second message certainly does not. By definition 5.2.1, the goal of Nb confidentiality is not available to B, who can never check his peer's activity. Intuitively, non-availability of the goal signifies that the existing guarantees of Nb confidentiality are of no use to B. Not surprisingly, when the assumption about A's activity is not verified in practice, Nb indeed loses confidentiality, and then the Spy can mount the known man-in-the-middle attack.

Lowe's corrected protocol mentions B in the second message. An analogous reasoning for this protocol is given in Theorem 5.4.2. Here, *evs* is a generic trace of the model for the corrected protocol, which can be found with file NS_Public.thy [33, 34].

Theorem 5.4.2 (NSPL_Spy_not_see_NB). *If A and B are uncompromised and evs contains*

Says B A (Crypt(pubK A){|Nonce Na, Nonce Nb, Agent B|})

then evs is such that Nonce Nb $\notin$ analz(spies evs).

This theorem confirms that the corrected protocol makes Nb confidentiality available to B, who is guaranteed within his minimal trust that his nonce is kept secret (and Lowe's attack cannot succeed).

The details of these protocols were known before the present research. However, we believe that studying availability of the confidentiality goal to B would have helped to discover the lack of explicitness in the second message. Faced with Theorem 5.4.1, the human analyser would have pragmatically checked if any extra explicitness reduced B's assumptions to either verifiable ones or to minimal trust assumptions. This is exactly how we will proceed in general. Without goal availability, it is difficult to evaluate whether Theorem 5.4.1 as it stands conveys a satisfactory guarantee.

We will show that studying availability of the confidentiality goal on the Shoup-Rubin protocol highlights a previously unnoticed lack of explicitness

(Chapter 11). These considerations support the claim that non-availability of a confidentiality goal indicates lack of explicitness. However, the claim does not always hold. We will show on Kerberos IV (Chapter 7) that confidentiality of a session key is not available to the protocol responder in a realistic model, although this is not due to lack of explicitness. Therefore, studying availability of other goals may also be important, as clarified below on the original Otway-Rees protocol (§8.5.1). An assessment on availability of the key distribution goal highlights that this goal can be made available to the initiator by merely strengthening one protocol message. This finding undermines the BAN logic claim [58] that, if a session key is never used to encrypt a protocol message, then no agent is entitled to know that his peer knows that key.

6. Modelling Timestamping and Verifying a Classical Protocol

The Inductive Method is extended with the treatment of timestamps, which, contrarily to nonces, are guessable numbers. Then, the first timestamp-based protocol, BAN Kerberos, is modelled and verified while serving to demonstrate the extensions.

The original Inductive Method does not include timestamps among the formalised message components (Chapter 3) and is in fact only benchmarked on nonce-based protocols such as Otway-Rees and Yahalom.

At the beginning of the 1980s, Denning and Sacco pioneered the use of timestamps in the field of security protocols [71] to avoid replay attacks. Timestamps, which are numbers marking a specific instance of time, have been employed since then in many protocols such as BAN Kerberos [58, §6] and Kerberos IV [121]. We extend the Inductive Method with the treatment of timestamps in order to analyse this new class of protocols.

The single operational difference between nonces and timestamps is that only the latter can be guessed by the Spy. So, we model timestamps as guessable numbers to include in the allowed message components. The price is limited to that of introducing the new message constructor, proving a few technical lemmas, and doing minor updates to some of the existing ones. The approach becomes significantly more general. The new message component is also used to model any extra information that the real-world protocols pass inside their messages, which is in general available to the Spy. This is the case, for example, with the session identifier and other fields of the model for the TLS protocol [134].

The first benchmark we choose for the extended approach is the BAN Kerberos protocol, as its well-known analysis by BAN logic [58, §6] provides a significant opportunity for comparison. The BAN analysis concludes that, if A has completed a session of the protocol with B, then A is aware that B meant to communicate with A using the session key Kab; the equivalent guarantee is offered to B, resulting in

$$A \models B \models A \xleftrightarrow{Kab} B \quad \text{and} \quad B \models A \models A \xleftrightarrow{Kab} B$$

These conclusions may be viewed as mutual non-injective agreement on the session key. Our findings confirm this goal (via some extensions to the approach, see §8.3) and strengthen it with a deeper investigation of others such as authenticity and confidentiality [41]. The proofs, partially adapted from the existing ones on the shared-key Needham-Schroeder protocol, also suggest refining the treatment of the accidental losses of session keys, the oops events. Our protocol model also accounts for the temporal checks performed by the agents at each step. This helps us explain how the protocol functions beyond the mere sequencing of messages.

The Kerberos project started at MIT during the mid 1980s [121] and, over a decade, generated several variants of the same protocol design [102]. BAN Kerberos is considered the natural modification of the shared-key Needham-Schroeder protocol with the addition of timestamps. Therefore, it is interesting to compare the temporal requisites that the two protocols add to the goal of authentication. This is achieved later (§8.6) because the Inductive Method must be suitably extended.

This chapter presents our formalisation of guessable numbers (§6.1) and of a discrete time (§6.2), which were released with the 1999 distribution of Isabelle [158]. Then, it introduces the BAN Kerberos protocol (§6.3), its inductive model (§6.4) and the corresponding guarantees (§6.5).

6.1 Modelling Guessable Numbers

Guessable numbers are inherently different from nonces: all guessable numbers are assumed to be known to the Spy. For example, they are used to mark time instances or to specify message options. Conversely, nonces are long random numbers that are extremely difficult to guess, and our model assumes them to be impossible to guess (§3.10).

In consequence, guessable numbers must be modelled separately from nonces. This requires three simple steps. First, the Isabelle datatype msg is extended with a new constructor Number that takes as its parameter a natural number. A timestamp T will be represented in the model by the message component Number T.

Then, the inductive definition of synth H seen above (§3.10) is extended with a fifth rule to allow the Spy to synthesise any number; this is the major difference with nonces.

5. Any number can be synthesised from any message set.

  Number $N \in$ synth H

Finally, the main operators need suitable rewriting rules for the symbolic evaluations that involve the new component. They are easy to obtain as two technical lemmas.

$$\text{parts}(\{\text{Number } N\} \cup H) \;=\; \{\text{Number } N\} \cup (\text{parts } H)$$
$$\text{analz}(\{\text{Number } N\} \cup H) \;=\; \{\text{Number } N\} \cup (\text{analz } H)$$

Minor updates necessary to a few existing lemmas are omitted here. Once guessable numbers are modelled, a formal treatment of time can be conceived, as we shall see in the next section.

6.2 Modelling Time

Traces only grow linearly. If the event ev is located after the event ev' in a trace, then ev' in the real world occurred at some later, unspecified, time after ev did. The trace model assumes that no two events occur simultaneously, an assumption that can be relaxed by defining a trace as a list of sets of events. However, this does not appear to be necessary. If $ev1$ and $ev2$ really occurred concurrently, the protocol model contains a trace where $ev1$ precedes $ev2$ and another trace where $ev2$ precedes $ev1$. Concurrency is therefore modelled as the two corresponding sequential approximations.

If we imagine that each event in a trace carries the time instant when it occurred, then a trace provides a sampling of time. This corresponds to defining an injective function between the set of all traces and the set of all samplings of time. Therefore, the time sampling associated with a trace is such that the first value of the sampling represents the time when the first event in the trace took place, the second value represents the time for the second event in the trace; and so on. The function is clearly not a bijection because there may exist different traces corresponding to the same sampling. The empty sampling is associated with the empty trace.

For simplicity, we can *normalise* a time sampling in terms of segments of natural numbers as follows. Let the segment 0 correspond to the empty sampling. Let the segment 1 correspond to any sampling containing only one value; let the segment 1, 2 correspond to any sampling containing only two values; let the segment 1, 2, 3 correspond to any sampling containing only three values, and so on. This indeed is a simplification as it hides the precise temporal relations between events occurring on different traces. However, these appear to be irrelevant to the experiments we discuss in the following.

Since the current time of a trace is the highest value of the corresponding time sampling, after normalisation the current time of a trace of length n exactly is n. In consequence, we declare the function

CT : event list $\longrightarrow$ nat

and define

CT evs $\triangleq$ length evs

Recall that a trace of length n represents a history of the network during which n events have taken place. Then, it seems intuitive to think that the current time of the trace is precisely n. Clearly, this formalisation hides the problem of keeping remote clocks synchronised: each agent's minimal trust includes that his peers' clocks are synchronised with his own. The treatment presented here will be used in the rest of this book for the analyses of the protocols that make use of timestamps and of the associated notion of time.

6.3 The BAN Kerberos Protocol

BAN Kerberos [58, §6] is a key distribution protocol, namely it aims at distributing session keys to its peers (Figure 6.1).

$$
\begin{array}{llll}
1. & A & \rightarrow & S & : & A, B \\
2. & S & \rightarrow & A & : & \{Tk, B, Kab, \underbrace{\{Tk, A, Kab\}_{Kb}}_{ticket}\}_{Ka} \\
3. & A & \rightarrow & B & : & \underbrace{\{Tk, A, Kab\}_{Kb}}_{ticket}, \underbrace{\{A, Ta\}_{Kab}}_{authenticator} \\
4. & B & \rightarrow & A & : & \{Ta + 1\}_{Kab}
\end{array}
$$

Fig. 6.1. BAN Kerberos protocol

The design closely resembles that of the shared-key Needham-Schroeder protocol (§2.2.4) but rests on a different procedure of mutual authentication (in particular the last two steps). The protocol associates one lifetime with session keys and another with authenticators. Lifetimes represent the time intervals within which the corresponding components should be considered valid. The lifetimes are passed in the messages but we omit them from the presentation, assuming they are known to all.

After A's initial request to establish a session with B, the Server issues a fresh session key and includes it, along with the timestamp that marks its time of issue, inside a message sealed with A's shared key. The message also contains a ticket sealed with B's shared key, which in turn contains a duplicate of the session key and its timestamp. The message is sent to A, who removes the external encryption and learns that the session key Kab issued at time Tk is indeed meant for the session with B. Then, A checks that Tk has not expired to establish whether the session key is still valid. If so, A builds an authenticator with a new timestamp Ta and sends it with the ticket to B. Upon reception of this message, B decrypts the ticket and learns the session key for the session with A and its timestamp Tk. If Tk has not expired, B uses the session key to decipher the authenticator. This should give him evidence that A was alive and able to use the session key at time Ta. The same guarantee should be given to A by the certificate that B sends her in the final step of the protocol. However, the choice of $Ta + 1$ is imprecise because it is not necessarily the time when B acts. It might be the historical influence of the shared-key Needham-Schroeder protocol to have determined this design choice, although BAN Kerberos uses timestamps rather than nonces. It would be preferable that B insert the current time.

6.4 Modelling BAN Kerberos

This protocol uses the notion of time in two ways. One is the issue of timestamps as the current time. Having defined a simple model of discrete time in the previous section, we can use it here.

Another use of time is the checking of timestamps against the current time with respect to specific lifetimes. For this purpose, we declare two natural numbers, sesKlife and authlife, to formalise respectively the lifetimes of session keys and authenticators. Agents check the timestamps against them and discard the messages containing expired timestamps. To formalise those checks, we declare two binary predicates

expiredK, expiredA : [nat, event list]

and define them as

expiredK Tk evs $\triangleq$ (CT evs) $-$ Tk $>$ sesKlife

expiredA Ta evs $\triangleq$ (CT evs) $-$ Ta $>$ authlife

When, for example, expiredK Tk evs holds, a longer time than sesKlife has elapsed since Tk at the moment when the history recorded by evs is examined. Therefore, the session key associated with Tk is no longer valid on evs (which technically has become too long). The association between Tk and its session key is established by the structure of the second protocol message, so it does not need explicit formalisation.

The constant bankerberos, declared as a set of lists of events, represents the formal protocol model and is defined by induction in Figure 6.2. It can be found with file Kerberos_BAN.thy (Figure 3.1). The empty trace formalises the initial scenario, in which no protocol session has taken place. Rule *Nil* sets the base of the induction stating that the empty trace is admissible in the protocol model. All other rules represent inductive steps, so they detail how to extend a given trace of the model.

Rule *Fake* models the Spy's illegal activity, which includes forging any timestamps. Rule *BK1* lets any agent begin a protocol session at any time. Rule *BK2* models the Server's operation, which is subordinate to some other agent's having sent the first message of the protocol. Since the first message is a cleartext, the Spy may easily fake it many times and overload the Server. The session key has not been used before and is accompanied with a timestamp drawn from the current time. Rule *BK3* states that an agent who initiated a protocol session may proceed with the third step of the protocol if she has been sent a message with a non-expired session-key timestamp. By

Nil:
[] ∈ bankerberos

Fake:
⟦ evsF ∈ bankerberos; X ∈ synth (analz (spies evsF)) ⟧
⟹ Says Spy B X # evsF ∈ bankerberos

BK1:
⟦ evs1 ∈ bankerberos ⟧
⟹ Says A Server {|Agent A, Agent B|} # evs1 ∈ bankerberos

BK2:
⟦ evs2 ∈ bankerberos; Key Kab ∉ used evs2; Kab ∈ symKeys;
  Says A' Server {|Agent A, Agent B|} ∈ set evs2 ⟧
⟹ Says Server A (Crypt (shrK A) {|Number (CT evs2), Agent B, Key Kab,
                Crypt (shrK B) {|Number (CT evs2), Agent A, Key Kab|}|})
       # evs2 ∈ bankerberos

BK3:
⟦ evs3 ∈ bankerberos;
  Says A Server {|Agent A, Agent B|} ∈ set evs3;
  Says S A (Crypt (shrK A) {|Number Ts, Agent B, Key Kab, Ticket|})
    ∈ set evs3;
  ¬ expiredK Ts evs3 ⟧
⟹ Says A B {|Ticket, Crypt Kab {|Agent A, Number (CT evs3)|}|}
       # evs3 ∈ bankerberos

BK4:
⟦ evs4 ∈ bankerberos;
  Says A' B {|Crypt (shrK B) {|Number Ts, Agent B, Key Kab|},
              Crypt Kab {|Agent A, Number Ta|}|} ∈ set evs4;
  ¬ expiredK Ts evs4; ¬ expiredA Ta evs4 ⟧
⟹ Says B A (Crypt Kab (Number Ta))
       # evs4 ∈ bankerberos

Oops:
⟦ evs0 ∈ bankerberos;
  Says Server A (Crypt (shrK A) {|Number Ts, Agent B, Key Kab,
                                  Ticket|}) ∈ set evs0 ⟧
⟹ Notes Spy {|Number Ts, Key Kab|}
       # evs0 ∈ bankerberos

Fig. 6.2. Inductive model of BAN Kerberos

rule *BK4*, an agent completes the protocol if he has been sent a specific intelligible message containing two non-expired timestamps. Observe that the model does not need to increment the timestamp in the last protocol step as the message is already structurally different from all others. Finally, rule *Oops* as it stands allows accidental leaks of session keys at any time. It will be refined (§6.6) to only allow leaks of keys that have expired.

Observe that rules *BK1* to *BK4* model the agents' behaviour that is legal according to BAN Kerberos. For example, it is visible that only the correct timestamps are inserted. Even the Spy can execute them, a feature that models her legal behaviour.

6.5 Verifying BAN Kerberos

The main guarantees that we have proved about BAN Kerberos are presented in this section, where *evs* always stands for a generic trace of the formal protocol model **bankerberos**. The authentication goals (§6.5.6) will be subsequently strengthened thanks to a few extensions to our approach (§8.3). The theorem names follow our naming conventions (§1.3.2).

6.5.1 Reliability of the BAN Kerberos Model

The only relevant reliability theorem states that the model Server is reliable (Theorem 6.5.1). If the certificate is addressed to A, then the Server encrypts it with A's shared key, and inserts a session key and the ticket meant for A's peer, B. Since cryptographic keys can be either long-term keys or session keys, Kab is shown not to be a long-term one; since it is never used before the Server issues it, it is fresh. The timestamp is chosen as the current time of the subtrace where the key is being issued.

Theorem 6.5.1 (BK_Says_Server_message_form). *If evs contains*

$ev = $ Says Server A (Crypt K {Number Tk, Agent B, Key Kab, $Ticket$})

then

$K = $ shrK A *and* $Kab \in $ symKeys *and* $Kab \notin $ range shrK *and*
$Ticket = $ Crypt(shrK B) {Number Tk, Agent A, Key Kab}

and evs is such that

Key $Kab \notin $ used(before ev on evs) *and*
$Tk = $ CT(before ev on evs).

Proving the last two conjuncts of the assertion requires some minor extensions to the classical strategy (§4.1). Various lemmas must be applied for the symbolic evaluation of the length of a subtrace, as is required by the before function.

6.5.2 Regularity

The protocol employs a single kind of long-term key and never sends it in the traffic, so a key regularity lemma is provable. An agent's shared key can be analysed from the traffic if and only if the agent is compromised (Lemma 6.5.1).

Lemma 6.5.1 (BK_Spy_analz_shrK). *Trace evs is such that* Key(shrK A) $\in$ analz(spies *evs*) *if and only if* $A \in$ bad.

All subsequent guarantees about certificates sealed with a shared key will appeal to this lemma to guarantee their integrity.

6.5.3 Authenticity

The second message and the ticket contained inside it represent the crucial certificates of the protocol because they are meant to deliver the session key to A and B respectively. They are encrypted under shared keys. Applying the regularity lemma assures that the Spy cannot handle the keys that encrypt the certificates, and so cannot spoof them.

Let us consider the certificate for A and assume the agent to be uncompromised in order to apply the regularity lemma. When A receives the certificate and inspects it, she realises that it is the four-component certificate delivering a session key. The theorem proves that the certificate was created by the only entity that is legally entitled to issue session keys, the Server; so it is authentic (Theorem 6.5.2).

Theorem 6.5.2 (BK_Kab_authentic). *If A is uncompromised and evs is such that*

Crypt(shrK A){Number Tk, Agent B, Key Kab, $Ticket$} $\in$ parts(spies *evs*)

then evs contains

Says Server A (Crypt(shrK A){Number Tk, Agent B, Key Kab, $Ticket$}).

Since A is uncompromised, the certificate is, via the regularity lemma, integral. Therefore, A learns that the session key was issued at time Tk. Checking this timestamp against the current time prevents her from accepting an old key as fresh. A similar guarantee [39] holds for the certificate {Na, B, Kab, X}$_{Ka}$ of the shared-key Needham-Schroeder protocol (Figure 2.6). Since the nonce Na was previously issued by A and then received along with the session key inside the integral certificate, A infers that the session key is more recent than her nonce.

An analogous guarantee can be established about the certificate for B (Theorem 6.5.3).

Theorem 6.5.3 (BK_ticket_authentic). *If B is uncompromised and evs is such that*

Crypt(shrK B){Number Tk, Agent A, Key Kab} $\in$ parts(spies evs)

then evs contains

Says Server A (Crypt(shrKA){Number Tk, Agent B, Key Kab,

Crypt(shrK B){Number Tk, Agent A, Key Kab}}).

Agent B is guaranteed that the session key was issued at time Tk, and so can verify its freshness. The corresponding guarantee for the shared-key Needham-Schroeder protocol [39] relies on the certificate ${K, A}_{Kb}$. This time B cannot decide whether he is accepting an old session key as fresh, an uncertainty that raises the known chance of a replay attack.

6.5.4 Unicity

The Server issues fresh session keys, so the same key cannot be issued twice. Precisely, we can enforce that, if a session key appears within two message contexts, then the contexts must be the same (Theorem 6.5.4).

Theorem 6.5.4 (BK_unique_session_keys). *If evs contains*

Says Server A (Crypt(shrK A){Number Tk, Agent B, Key Kab, $Ticket$})

and

Says Server A' (Crypt(shrK A'){Number Tk', Agent B', Key Kab, $Ticket'$})

then

$A = A'$ *and* $Tk = Tk'$ *and* $B = B'$ *and* $Ticket = Ticket'$.

This result is often used when proving theorems that assume the event that issues the session key. The *Oops* rule of the protocol model introduces another event of the same form, but they can be derived to be identical if they contain the same session key. A similar result could be enforced if two tickets containing the same, confidential, session key appear in the traffic.

While Theorem 6.5.4 allows the Server to send two identical messages, we can prove that this is impossible (Theorem 6.5.5). This is a stronger guarantee because it states that the Server never issues the same session key more than once for the same peers using the same timestamp and the same ticket.

Theorem 6.5.5 (BK_Server_Unique). *If evs contains*

Says Server A (Crypt(shrK A){Number Tk, Agent B, Key Kab, $Ticket$})

then

Unique (Says Server A (Crypt(shrK A)

{Number Tk, Agent B, Key Kab, $Ticket$})) on evs.

6.5.5 Confidentiality

The session key compromise theorem (**BK_analz_insert_freshK**, omitted here), which provides a crucial rewriting rule for the analz operator, can be proved conventionally since BAN Kerberos does not use session keys to encrypt other keys.

If the peers' shared keys are uncompromised, then no protocol step reveals to the Spy the session key that the Server issues. Moreover, if the key is not leaked by any accidents, then it can be proved to be confidential (Theorem 6.5.6).

Theorem 6.5.6 (BK_Confidentiality_S). *If A and B are uncompromised and evs contains*

Says Server A (Crypt(shrK A){Number Tk, Agent B, Key Kab, $Ticket$})

but does not contain Notes Spy {Number Tk, Key Kab}, *then evs is such that*

Key $Kab \notin$ analz(spies evs).

Even the application of the authenticity Theorem 6.5.2 still leaves the strong assumption that no oops event occurred involving the session key (Theorem 6.5.7), which we will relax later (§6.6).

Theorem 6.5.7 (BK_Confidentiality_A). *If A and B are uncompromised and evs is such that*

Crypt(shrK A){Number Tk, Agent B, Key Kab, $Ticket$} $\in$ parts(spies evs)

but does not contain Notes Spy {Number Tk, Key Kab}, *then evs is such that*

Key $Kab \notin$ analz(spies evs).

Likewise, session key confidentiality can be proved from B's viewpoint (Theorem 6.5.8) by application of the authenticity Theorem 6.5.3 to the confidentiality Theorem 6.5.6.

Theorem 6.5.8 (BK_Confidentiality_B). *If A and B are uncompromised and evs is such that*

Crypt(shrK B){Number Tk, Agent A, Key Kab} $\in$ parts(spies evs)

but does not contain Notes Spy {Number Tk, Key Kab}, *then evs is such that*

Key $Kab \notin$ analz(spies evs).

6.5.6 Authentication

One of the aims of BAN Kerberos is to enforce mutual agent authentication. The authenticator of the third message should authenticate A to B, and the fourth message as a whole should authenticate B to A. This section shows that the protocol establishes mutual weak agreement, while a few extensions to the approach will show that it also establishes mutual non-injective agreement on the session key (§8.3).

Tracing back the originator of the authenticator requires the session key that seals it to be confidential. Thus, the assumptions of the confidentiality Theorem 6.5.8 must be allowed, since we are reasoning from B's viewpoint. In these circumstances, the authenticator can be proved to have originated with A during the third step of the protocol (Theorem 6.5.9). Recall that the suffix of the theorem name indicates that the confidentiality assumption was relaxed by the appropriate formal argument.

Theorem 6.5.9 (BK_B_authenticates_A_r). *If A and B are uncompromised and evs is such that*

Crypt(shrK B){Number Tk, Agent A, Key Kab} $\in$ parts(spies *evs*)   *and*

Crypt Kab{Agent A, Number Ta} $\in$ parts(spies *evs*)

and evs does not contain Notes Spy {Number Tk, Key Kab}, *then it contains*

Says $A\,B$ {Crypt(shrK B){Number Tk, Agent A, Key Kab},

Crypt Kab{Agent A, Number Ta}}.

This theorem is useful to B. It says that, if B receives the ticket, extracts the session key, and then receives the authenticator that is sealed with it, he can infer something useful: A was alive and meant to communicate with him by sending him the concatenated message. The theorem does not express A's knowledge of Kab or Ta.

The same strategy proves that a certificate that has the form of the fourth message of the protocol was indeed created during the fourth step (Theorem 6.5.10). The necessary condition that the certificate be sealed with a confidential session key is conveyed via an appeal to the confidentiality Theorem 6.5.7. As usual, the suffix of the theorem name indicates that the confidentiality assumption was relaxed by the appropriate formal argument.

Theorem 6.5.10 (BK_A_authenticates_B_r). *If A and B are uncompromised and evs is such that*

Crypt(shrK A){Number Tk, Agent B, Key Kab, $Ticket$}

$\in$ parts(spies *evs*)   *and*

Crypt Kab(Number Ta) $\in$ parts(spies *evs*)

and evs does not contain Notes Spy {Number Tk, Key Kab}, *then it contains*

Says $B\,A$ (Crypt Kab(Number Ta)).

This theorem is useful to A: when A gets hold of the two suitable certificates, she infers that B was alive and meant to communicate with her. The theorem does not express B's knowledge of Kab or Ta.

6.5.7 Key Distribution

We want to establish whether, at the end of a protocol session, the peers have evidence that they share a session key, which is a major goal of the protocol.

In order to decrypt the authenticator $\{\!|A, Ta|\!\}_{Kab}$, agent B must learn Kab from the ticket $\{\!|Tk, A, Kab|\!\}_{Kb}$. Under the strong assumption that the session key was not leaked by accident, B appeals to the confidentiality argument and concludes that Kab is confidential. Therefore, the authenticator cannot have been faked and, so, must have been sent by A in the third step of the protocol. The authentication Theorem 6.5.9 establishes this formally. Also, since A is the true creator of the authenticator, she must know the session key to seal it. This notion cannot be captured formally in the current approach: in general, when an event $\mathsf{Says}\, A\, B\, X$ occurs, A may just be forwarding X and therefore have no knowledge about its contents.

To overcome this limitation, we will extend the Inductive Method in Chapter 8 to capture the notion of an agent's being the true creator of a message. Modelling message reception will provide an alternative reasoning: when B receives the ticket and the authenticator, A must have previously received an instance of the second message and so must have learnt the session key.

The authentication Theorem 6.5.10 allows the same considerations from A's viewpoint. Although the certificate $\{\!|Ta|\!\}_{Kab}$ is proved to have been sent by B, proving B's knowledge of the session key requires, as mentioned above, further modelling.

6.6 A Temporal Modelling of Accidents

All confidentiality and authentication theorems require the condition that no oops event occurred, but no agent other than the Spy can verify this in practice. Can this condition be considered part of the agents' minimal trust? An affirmative answer seems reasonable because accidents are always unforseen. In consequence, the protocol makes session key confidentiality and authentication available to the peers (§2.2.4 and Chapter 5).

This delicate point raises concern that allowing the leaking of any session key at any time may produce an overly pessimistic threat model. We observe that the longer a message component is in the traffic, the higher is the risk that the Spy may get hold of it. In particular, the probability that a session key becomes compromised increases over time.

In light of these considerations, we can refine the threat model by introducing a temporal modelling of accidents. We assume that a session key cannot be leaked as long as its lifetime has not expired, namely the key cannot be leaked as long as it is still valid. Incorporating the change in the protocol model requires adding the precondition

expiredK Tk $evs O$

to the *Oops* rule. Hence, session-key leaks may only happen to histories that have evolved for longer than sesKlife after the time of issue of the session key. In the new model, all confidentiality and authentication theorems (6.5.6 to 6.5.10) can be refined so as to rest on the assumption ¬expiredK Tk evs, namely "Tk has not expired on evs," instead of "evs does not contain Notes Spy {Number T, Key Kab} for any T." They are omitted here but can be found in the repository [33, 34]. Their names have suffix _**temporal**.

Any agent can check the new temporal assumption by verifying that the current time does not differ from the timestamp by more than the allowed lifetime. Therefore, all confidentiality and authentication guarantees are applicable by (and the corresponding goals available to) their respective beneficiaries in new the threat model. This further confirms that studying adherence to a principle of protocol design or analysis is subject to the assumed threat model. Our conclusions are the same under the two threat models but are easier to derive under the temporal modelling of accidents, provided that each agent's minimal trust has all the network clocks synchronised.

The temporal modelling of accidents will be adopted throughout this book for the analysis of those protocols that make use of timestamps: Kerberos IV (Chapter 7) and Kerberos V (Chapter 9).

7. Verifying a Deployed Protocol

Kerberos IV is the first deployed protocol to be modelled and verified using the Inductive Method. A weakness is discovered in its management of timestamps, leading to realistic confidentiality attacks on a class of session keys and to corresponding authentication attacks.

The most commonly deployed variant of the Kerberos protocol is Version IV [121]. Kerberos IV is a password-based system for authentication and authorisation over local area networks. It was developed during the late 1980s with the aim of implementing a robust strategy for single access: once a user authenticates himself to a network machine, the process of obtaining authorisation to access a network service should be completely transparent to him. Hence, the user should only have to enter his password once during the authentication phase, and never during the subsequent authorisation phases. Kerberos IV pursues its aim by delivering certain credentials to the login process of each user during the authentication phase. These credentials are to be used with a suitable trusted Server during any subsequent authorisation phase. Receiving, storing and using the credentials are transparent to the user.

Both trusted Servers of the authentication phase and of the authorisation phase are modelled below. Each of them issues a session key that is to be considered valid within a specific lifetime. The session key issued by the first Server is used to encrypt the one issued by the second Server. This feature greatly complicates the verification of the confidentiality goals [40], but a technique for modelling the association between two message components can be reused from the analysis of the Yahalom protocol [135]. The threat model adopted for this analysis is conventional Dolev-Yao's extended with agents' compromises (§3.9) and with the temporal modelling of accidents introduced on BAN Kerberos (§6.6). First, the authenticity, unicity, confidentiality, authentication and key distribution goals are investigated. Then, in verifying adherence to our principle of goal availability, we discover a weakness in the management of timestamps. In our threat model, the weakness leads to a simple confidentiality attack on a session key, which is then used to mount an authentication attack: the Spy may access a network service using a session key that belongs to a user who is no longer present on the network. Our proofs suggest a simple fix, which can be formally verified to be effective.

To our knowledge, this is the first mechanised proof of correctness for the complete protocol, although some related work exists. Mitchell et al. [123] model check a highly simplified version of the protocol, which derives from Kohl et al. [101]. Neither are timestamps included in their model, nor are multiple runs allowed. They find no attacks on a system of size 3 — consisting of an initiator, the Kerberos Servers and a responder — and a "redirection" attack on a system of size four, including two responders, which the full Kerberos IV prevents by means of explicitness (§7.1.2).

We earlier analysed the protocol by ASMs (§2.1.1) with Riccobene [45], formalising the actions of an unbounded population of agents by means of a detailed algebraic model. An intuitive if-then-else language was used to formalise the protocol steps, but the Spy's illegal activity was defined *ad hoc* for the protocol. The present inductive analysis starts off from the ASMs' formal specification rather than from the informal technical report about the protocol [121]. This made the development of the inductive protocol model considerably faster than was expected. The human analyser's digestion of informal specifications may turn out to be very time consuming, as our subsequent work on the SET protocol would confirm [37].

This chapter describes the three phases of Kerberos IV (§7.1), its inductive modelling (§7.2) and its mechanised verification (§7.3).

7.1 The Kerberos IV Protocol

Kerberos IV is essentially a key distribution protocol. It relies on the Kerberos System (Figure 7.1), which comprises two trusted Servers and a database containing all users' passwords sealed using the standard Unix one-way encryption algorithm.

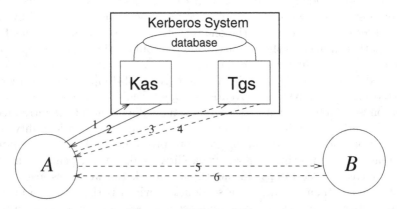

Fig. 7.1. Kerberos IV layout

7.1.1 Overview

Once a user types in his identifier and password, his login process seals the password using the Unix algorithm. This process is the agent who initiates the protocol (recall that we view agents as processes, §3.3). The full protocol consists of three phases, only the first being compulsory. The first phase, AUTHENTICATION, comprises the first two protocol steps and serves to authenticate the initiator to the Kerberos System, specifically to the Kerberos Authentication Server, Kas for short. If the user is registered, the two agents share the sealed password as a long-term secret, which constitutes the initiator's shared key: the initiator has computed it, while Kas has looked it up in its database. Using this secret, Kas issues some authorisation credentials that the initiator will use in the second phase, AUTHORISATION, which comprises the third and fourth protocol steps. This phase only occurs when the initiator, who is currently running on a workstation, requires a network service. Using the previously obtained authorisation credentials, the initiator contacts the Kerberos System, specifically the Ticket Granting Server, Tgs for short. The initiator obtains some service credentials to use in the final phase, SERVICE, in order to access the requested service.

Observe that, while Kas does not have a long-term key of its own, Tgs, like all other agents, does have one that is shared with Kas.

7.1.2 Details

The complete protocol is presented in Figure 7.2.

During the AUTHENTICATION phase, the initiator A queries Kas with her identity, Tgs and a timestamp $T1$; Kas issues a session key and looks up A's shared key in the database. It replies with a message sealed with A's shared key containing the session key, its timestamp Ta, Tgs and a ticket. The session key and the ticket are the credentials to use in the subsequent authorisation phase, so we address them as *authkey* and *authticket* respectively. The authticket is sealed with Tgs's shared key and contains a copy of the authkey, its timestamp and its peers. The lifetime of an authkey is several hours.

If $T1$ is not much older than Ta with respect to a given lifetime, then A is assured that Kas's reply was prompt. If this check is affirmative, A may want to start the AUTHORISATION phase. She sends Tgs a three-component message including the authticket, an authenticator sealed with the authkey containing her identity and a new timestamp $T2$, and B's identity. The lifetime of an authenticator is several seconds. Upon reception of the message, Tgs decrypts the authticket, extracts the authkey and checks the validity of its timestamp Ta, namely that Ta is not too old with respect to the lifetime of authkeys. Then, Tgs decrypts the authenticator using the authkey and checks the validity of $T2$ with respect to the lifetime of authenticators. Finally, Tgs issues a new session key and looks up B's shared key in the

AUTHENTICATION

1. $A \rightarrow$ Kas : $A, \mathsf{Tgs}, T1$

2. Kas $\rightarrow A$: $\{authK, \mathsf{Tgs}, Ta, \underbrace{\{A, \mathsf{Tgs}, authK, Ta\}_{Ktgs}}_{authTicket}\}_{Ka}$

AUTHORISATION

3. $A \rightarrow$ Tgs : $\underbrace{\{A, \mathsf{Tgs}, authK, Ta\}_{Ktgs}}_{authTicket}, \underbrace{\{A, T2\}_{authK}}_{authenticator}, B$

4. Tgs $\rightarrow A$: $\{servK, B, Ts, \underbrace{\{A, B, servK, Ts\}_{Kb}}_{servTicket}\}_{authK}$

SERVICE

5. $A \rightarrow B$: $\underbrace{\{A, B, servK, Ts\}_{Kb}}_{servTicket}, \underbrace{\{A, T3\}_{servK}}_{authenticator}$

6. $B \rightarrow A$: $\{T3 + 1\}_{servK}$

Fig. 7.2. Kerberos IV protocol

database. It replies with a message sealed with the authkey containing the new session key, its timestamp Ts, B and a ticket. The session key and the ticket are the credentials to use in the subsequent service phase, so we address them as *servkey* and *servticket* respectively. The servticket is sealed with B's shared key and contains a copy of the servkey, its timestamp and its peers. The lifetime of a servkey is a few minutes.

If $T2$ is not much older than Ts with respect to a given lifetime, then A is assured that Tgs's reply was prompt. If so, A may start the SERVICE phase. She sends B a two-component message including the servticket and an authenticator sealed with the servkey containing her identity and a new timestamp $T3$. Upon reception of the message, B decrypts the servticket, extracts the servkey and checks the validity of its timestamp Ts. Then, B decrypts the authenticator using the servkey and checks the validity of $T3$. Finally, B increments $T3$, seals it with the servkey, and sends it back to A.

In a simplified version of the protocol [101], the fourth message does not include either of the two occurrences of B's identity. The Spy can then mount the mentioned redirection attack [123]: she changes B's identity to some compromised C's (or her own) in the third message; she intercepts the fifth message, decrypts the servticket, extracts the servkey, decrypts the authenticator, and extracts the timestamp $T3$; she forges the sixth message. Hence, A believes to have completed a protocol session with B when in fact she has been redirected to someone else and B has never participated. The complete Kerberos IV clearly does not suffer this attack.

7.2 Modelling Kerberos IV

To model the two trusted Servers, we might modify the Isabelle datatype of agents (§3.3) as

datatype agent ≜ Kas
$\qquad\qquad\quad$ Tgs
$\qquad\qquad\quad$ Friend nat
$\qquad\qquad\quad$ Spy

Then, the definition of initState would have to be updated to allow both trusted Servers to know all agents' shared keys. Also, both Kas and Tgs would have to be assumed uncompromised. However, we get the same outcome from a less invasive formalisation, which merely sees the two Servers as translations

Kas ≜ Server

Tgs ≜ Friend 0

Friendly agents remain an infinite population. Because Server already is assumed uncompromised, we only need to state that so is Tgs.

The protocol reliance on time is identical to that of BAN Kerberos, so the same discrete formalisation of the current time of a trace (§6.4) is adopted here. We also need to define three natural numbers, authKlife, servKlife, authlife, formalising respectively the lifetimes of an authkey, a servkey, and an authenticator. Three intuitive binary predicates

expiredAK, expiredSK, expiredA : [nat, event list]

check their respective validities, namely the validities of the timestamps associated with them. They can be easily defined as

expiredAK Ta evs ≜ (CT evs) $-$ Ta > authKlife

expiredSK Ts evs ≜ (CT evs) $-$ Ts > servKlife

expiredA $\quad T \quad evs$ ≜ (CT evs) $- T \quad$ > authlife

When any of these predicates hold for some timestamp and some trace, we say that the timestamp has expired or that the corresponding item (a key or an authenticator) has expired on that trace.

A further lifetime, replylife, indicates the validity interval of any trusted Server's replies. A suitable binary predicate over natural numbers, defined as

valid T wrt T' ≜ $T \leq T' +$ replylife

checks that a timestamp is issued within such a lifetime after the instant marked by another timestamp. Agents operate this check to discard late Servers' replies, which may be due either to the Servers' temporary malfunction or to network latency. Conversely, those replies arriving within the specified lifetime are more reliable.

The constant kerbIV represents the formal protocol model and is defined by induction below. It is declared as a set of lists of events, and comes with the file KerberosIV.thy (Figure 3.1).

7.2.1 Basics

The inductive definition of kerbIV contains two basic rules (Figure 7.3). The first sets the base of the induction introducing the empty trace in the protocol model (Nil). The second models the Spy's illegal activity allowing the Spy to send any agent any fake message derived from her active analysis of the traffic (Fake).

> Nil:
> [] ∈ kerbIV
>
> Fake:
> ⟦ evsF ∈ kerbIV; X ∈ synth (analz (spies evsF)) ⟧
> ⟹ Says Spy B X # evsF ∈ kerbIV

Fig. 7.3. Inductive model of Kerberos IV: basics

7.2.2 Authentication Phase

The protocol initiator A must go through the authentication phase with Kas (Figure 7.4).

Any trace, including the empty one, can be extended by the event formalising the first message of the protocol (KIV1). This is faithful to the real world, where any agent, including the Spy, may decide to initiate a protocol session at any time. In the model, Kas's reply is subject to the previous occurrence of a suitable request from some agent (KIV2), whereas in the real world Kas does not operate until it actually receives the request. The model cannot be that precise at this stage because we have not yet formalised message reception (§8.2). It can be seen that Kas issues a fresh authkey, which is a symmetric key.

7.2.3 Authorisation Phase

The initiator A may require authorisation to a network service (Figure 7.5).

KIV1:
evs1 ∈ kerbIV
⟹ Says A Kas ⦃Agent A, Agent Tgs, Number (CT evs1)⦄
     # evs1 ∈ kerbIV

KIV2:
⟦ evs2 ∈ kerbIV; Key authK ∉ used evs2; authK ∈ symKeys;
   Says A' Kas ⦃Agent A, Agent B, Number T1⦄ ∈ set evs2 ⟧
⟹ Says Kas A (Crypt (shrK A) ⦃Key authK, Agent Tgs, Number (CT evs2),
                                Crypt (shrK Tgs) ⦃Agent A, Agent Tgs,
                                     Key authK, Number (CT evs2)⦄⦄)
        # evs2 ∈ kerbIV

Fig. 7.4. Inductive model of Kerberos IV: authentication phase

Before contacting Tgs, *A* checks that someone issued a message containing the authorisation credentials for her. While the message must have a specific form, its sender can be anyone, even the Spy. The reception of such a message is, as above, implicit (*KIV3*). Observe that *A* checks that the authkey was not issued too late after her request. She does not need to check that the authkey has not expired as this is the Kerberos Servers' concern (the check was stated in a previous formalisation [40]). Once a trace records a query is the expected form, Tgs may issue its reply, which contains a fresh servkey, provided that neither the authkey nor the authenticator have expired (*KIV4*).

7.2.4 Service Phase

This phase can be modelled (Figure 7.6) by the same layout as that used for the authorisation phase. The initiator *A* can contact the requested service *B* only if the trace records the issue of a servkey within the validity interval from *A*'s request. As with the authkey, *A* is not concerned that the servkey has not expired (*KIV5*). Before acknowledging *A*'s request, *B* checks that neither the servkey nor the authenticator have expired (*KIV6*). As with BAN Kerberos, we decide not to model the increment to the timestamp of the last step because it is imprecise: *B*'s reply may occur at a different time (§§7.3.6 and 8.4). From the verification standpoint, it matters that the last message is distinguishable by its form, regardless of the increment, from all others in the protocol.

7.2.5 Accidents

Authkeys are operationally different from servkeys, although they are all session keys. The two types of keys are associated with timestamps that are in turn associated with different lifetimes. Also, while authkeys are issued by

KIV3:

⟦ evs3 ∈ kerbIV;
   Says A Kas ⦃Agent A, Agent Tgs, Number T1⦄ ∈ set evs3;
   Says Kas' A (Crypt (shrK A) ⦃Key authK, Agent Tgs, Number Ta,
                                          authTicket⦄) ∈ set evs3;
   valid Ta wrt T1 ⟧
⟹ Says A Tgs ⦃authTicket, Crypt authK ⦃Agent A, Number (CT evs3)⦄,
               Agent B⦄
      # evs3 ∈ kerbIV

KIV4:

⟦ evs4 ∈ kerbIV; Key servK ∉ used evs4; B ≠ Tgs;
   authK ∈ symKeys; servK ∈ symKeys;
   Says A' Tgs ⦃Crypt (shrK Tgs) ⦃Agent A, Agent Tgs,
                                   Key authK, Number Ta⦄,
              Crypt authK ⦃Agent A, Number T2⦄, Agent B⦄ ∈ set evs4;
   ¬ expiredAK Ta evs4; ¬ expiredA T2 evs4 ⟧
⟹ Says Tgs A (Crypt authK ⦃Key servK, Agent B, Number (CT evs4),
                            Crypt (shrK B) ⦃Agent A, Agent B,
                            Key servK, Number (CT evs4)⦄⦄)
      # evs4 ∈ kerbIV

Fig. 7.5. Inductive model of Kerberos IV: authorisation phase

Kas and are meant to be used with Tgs, servkeys are issued by Tgs and are meant to be used with the network services. Furthermore, when an authkey expires, the corresponding user is logged out from the workstation and all his processes are killed. By contrast, when a servkey expires, the intended network service will no longer accept it, so the initiator must undertake a new authorisation phase.

Our threat model allows for accidental leaks of session keys (Figure 7.7) using the temporal modelling of accidents seen above (§6.6). An authkey can be noted by the Spy together with its timestamp and its peers provided that it has expired (*OopsA*), as can a servkey (*OopsS*). It is interesting to investigate whether the Spy can exploit expired session keys to get hold of non-expired ones, namely whether the attack cascades. This would be a major success for her, since the model allows the reuse of a session key within its lifetime, as the real world does.

7.3 Verifying Kerberos IV

Kerberos IV makes a peculiar use of session keys: Tgs employs an authkey to encrypt a servkey in the fourth message. This feature greatly complicates the verification of the confidentiality goals. Our proofs will show that syn-

KIV5 :
⟦ evs5 ∈ kerbIV; authK ∈ symKeys; servK ∈ symKeys;
  Says A Tgs {|authTicket, Crypt authK {|Agent A, Number T2|}, Agent B|}
    ∈ set evs5;
  Says Tgs' A (Crypt authK {|Key servK, Agent B, Number Ts,
                                servTicket|}) ∈ set evs5;
  valid Ts wrt T2 ⟧
⟹ Says A B {|servTicket, Crypt servK {|Agent A, Number (CT evs5)|}|}
    # evs5 ∈ kerbIV

KIV6 :
⟦ evs6 ∈ kerbIV;
  Says A' B {|Crypt (shrK B) {|Agent A, Agent B, Key servK, Number Ts|},
           Crypt servK {|Agent A, Number T3|}|} ∈ set evs6;
  ¬ expiredSK Ts evs6; ¬ expiredA T3 evs6 ⟧
⟹ Says B A (Crypt servK (Number T3)) # evs6 ∈ kerbIV

Fig. 7.6. Inductive model of Kerberos IV: service phase

OopsA :
⟦ evs0a ∈ kerbIV;
  Says Kas A (Crypt (shrK A) {|Key authK, Agent Tgs, Number Ta,
                                authTicket|}) ∈ set evs0a;
  expiredAK Ta evs0a ⟧
⟹ Notes Spy {|Agent A, Agent Tgs, Number Ta, Key authK|}
    # evs0a ∈ kerbIV

OopsS :
⟦ evs0s ∈ kerbIV;
  Says Tgs A (Crypt authK {|Key servK, Agent B, Number Ts, servTicket|})
    ∈ set evs0s;
  expiredSK Ts evs0s ⟧
⟹ Notes Spy {|Agent A, Agent B, Number Ts, Key servK|}
    # evs0s ∈ kerbIV

Fig. 7.7. Inductive model of Kerberos IV: accidents

chronising the issuing times of the two types of keys is crucial to making the
goal of servkey confidentiality available to the protocol responder.

   This section introduces suitable abbreviations to distinguish the two types
of session keys or to express the association between authkeys and servkeys,
and then presents all goals we verified of Kerberos IV. In the following, *evs* is
a generic trace of the set kerbIV. The theorem names follow the usual naming
conventions (§1.3.2).

7.3.1 Reliability of the Kerberos IV Model

Fragments of the proof script pertaining to the following guarantees can be found in Appendix A.1. To address the authkeys formally, we declare the function

authKeys : event list $\longrightarrow$ key set

so that authKeys evs yields all session keys that Kas issues in the trace evs, ignoring any eventual repetitions. Its definition is

authKeys evs $\triangleq$
    $\{authK \mid \exists\, A\ Ts \mid$
        Says Kas A (Crypt(shrK A){|Key $authK$, Agent Tgs, Number Ts,
                        Crypt(shrK Tgs){|Agent A, Agent Tgs,
                                    Key $authK$, Number Ts|}|})
      $\in$ set $evs\}$

Several lemmas are needed for the symbolic evaluation of this function. For example, one states formally that no authkeys appear on an empty trace (Lemma 7.3.1), and another signifies that Kas introduces an authkey (Lemma 7.3.2).

Lemma 7.3.1 (KIV_authKeys_empty). authKeys $[\,] = \{\}$.

Lemma 7.3.2 (KIV_authKeys_insert). *Trace evs is such that*

    authKeys(Says Kas A (Crypt(shrK A){|Key $authK$, Agent Tgs, Number Ta,
                              $authTicket$|})# evs)
    $= \{authK\} \cup$ (authKeys evs).

    Once a formalisation for the authkeys is available, the servkeys in a trace evs can be formalised as those keys that do not belong either to the range of shrK (so they are session keys) or to authKeys evs (so they are not authkeys).

    Our model Kas can be proved to be reliable (Theorem 7.3.1). When addressing a message to an agent A, Kas seals it with A's shared key and includes in it a session key that is an authkey along with a well-formed authticket. Using the function before (defined in §4.1), we establish that the authkey is fresh when it is issued and its timestamp is chosen as the current time.

Theorem 7.3.1 (KIV_Says_Kas_message_form). *If evs contains*

    $ev =$
    Says Kas A (Crypt K {|Key $authK$, Agent $Peer$, Number Ta, $authTicket$|})

then

$K = \mathsf{shrK}\,A$ *and* $Peer = \mathsf{Tgs}$ *and*
$authK \in \mathsf{symKeys}$ *and* $authK \notin \mathsf{range\,shrK}$ *and*
$authTicket = \mathsf{Crypt}(\mathsf{shrK\,Tgs})\{\!|\mathsf{Agent}\,A, \mathsf{Agent}\,\mathsf{Tgs},$
$\hspace{5cm}\mathsf{Key}\,authK, \mathsf{Number}\,Ta|\!\}$

and evs is such that

$authK \in \mathsf{authKeys}\,evs$ *and*
$\mathsf{Key}\,authK \notin \mathsf{used}(\mathsf{before}\,ev\,\mathsf{on}\,evs)$ *and*
$Ta = \mathsf{CT}(\mathsf{before}\,ev\,\mathsf{on}\,evs).$

An analogous guarantee enforces the reliability of the model Tgs (Theorem 7.3.2). An authkey is used to seal the message, which includes a fresh servkey and the servticket that quotes it.

Theorem 7.3.2 (KIV_Says_Tgs_message_form). *If evs contains*

$ev =$
Says Tgs A (Crypt $authK\,\{\!|\mathsf{Key}\,servK, \mathsf{Agent}\,B, \mathsf{Number}\,Ts, servTicket|\!\}$)

then

B *is not* Tgs   *and*
$authK \in \mathsf{symKeys}$ *and* $authK \notin \mathsf{range\,shrK}$ *and*
$servK \in \mathsf{symKeys}$ *and* $servK \notin \mathsf{range\,shrK}$ *and*
$servTicket = \mathsf{Crypt}(\mathsf{shrK}\,B)\{\!|\mathsf{Agent}\,A, \mathsf{Agent}\,B,$
$\hspace{5cm}\mathsf{Key}\,servK, \mathsf{Number}\,Ts|\!\}$

and evs is such that

$authK \in \mathsf{authKeys}\,evs$ *and* $servK \notin \mathsf{authKeys}\,evs$ *and*
$\mathsf{Key}\,servK \notin \mathsf{used}(\mathsf{before}\,ev\,\mathsf{on}\,evs)$ *and*
$Ts = \mathsf{CT}(\mathsf{before}\,ev\,\mathsf{on}\,evs).$

As can be expected, to obtain these two theorems, the general method for proving the reliability theorems (§4.1) must be enriched with frequent appeals to Lemmas 7.3.1 and 7.3.2.

7.3.2 Regularity

Kerberos IV never sends shared keys on the network, so a shared key is available to the Spy if and only if its owner is compromised. The corresponding regularity lemma has the usual formulation (Lemma 7.3.3).

Lemma 7.3.3 (KIV_Spy_analz_shrK). *Trace evs is such that*
$\mathsf{Key}(\mathsf{shrK}\,A) \in \mathsf{analz}(\mathsf{spies}\,evs)$ *if and only if* $A \in \mathsf{bad}$.

7.3.3 Authenticity

The protocol intends to deliver an authkey to A and Tgs and a servkey to A and B. Determining the originator of a certificate that contains a session key will confirm the authenticity of both the certificate and the key. As usual, we follow the principle of goal availability and perform the analysis from each agent's viewpoint. All proofs are carried out according to the methods described above (§4.3).

The instance of the second message that is sealed with A's shared key carries the authkey meant for A. An appeal to the regularity lemma guarantees that the certificate is tamperproof, so induction proves it to have originated with Kas (Theorem 7.3.3). Observe that the theorem becomes useful to A only upon reception of the certificate.

Theorem 7.3.3 (KIV_authK_authentic). *If A is uncompromised and evs is such that*

> Crypt(shrK A){|Key $authK$, Agent Tgs, Number Ta, $authTicket$|}
> $\in$ parts(spies evs)

then evs contains

> Says Kas A (Crypt(shrK A){|Key $authK$, Agent Tgs,
> Number Ta, $authTicket$|}).

The same guarantee can be enforced on the certificate that delivers the authkey to Tgs, the authticket (Theorem 7.3.4). Recall that Tgs is uncompromised.

Theorem 7.3.4 (KIV_authTicket_authentic). *If evs is such that*

> Crypt(shrK Tgs){|Agent A, Agent Tgs, Key $authK$, Number Ta|}
> $\in$ parts(spies evs)

then evs contains

> Says Kas A (Crypt(shrK A){|Key $authK$, Agent Tgs, Number Ta,
> Crypt(shrK Tgs){|Agent A, Agent Tgs, Key $authK$, Number Ta|}|}).

We now move on to investigating the authenticity of the servkey. The fourth message, which delivers it to A, is sealed with the authkey. Clearly, this must be A's authkey in order for the message to be intelligible to A. Since the authkey is not a shared key, the regularity lemma cannot help, so the authkey must be explicitly assumed to be confidential. The second and the fourth messages have the same structure. Theorem 7.3.3 pinpoints the second message by explicitly referring to a shared key as an encrypting key. By contrast, the fourth message uses a session key as an encrypting key, so this assumption must be made to investigate the authenticity of the message (Theorem 7.3.5). This version of the theorem is not applicable by A because of the confidentiality assumption on the authkey, which A cannot verify in practice. However, this assumption can be relaxed by the confidentiality argument (§7.3.5).

Theorem 7.3.5 (KIV_servK_authentic). *If evs is such that*

Crypt $authK$ {|Key $servK$, Agent B, Number Ts, $servTicket$|}
 ∈ parts(spies evs) *and*
Key $authK$ ∉ analz(spies evs) *and* $authK$ ∉ range shrK

then, for some A, evs contains

Says Tgs A (Crypt $authK$ {|Key $servK$, Agent B, Number Ts, $servTicket$|}).

As session keys typically differ in length from shared keys, it is fair to conclude that the authkey's not being a shared key is a verifiable assumption, and hence not a part of A's minimal trust. However, a variant of this theorem can be proved using another fact to discern between the second and the fourth messages: the latter states an agent that is certainly different from Tgs. So, the same conclusion holds by replacing the assumption on the authkey with B's not being Tgs (**KIV_servK_authentic_bis**, omitted here), which A can easily check by inspecting the fourth message. Another replacement is the entire event whereby Kas sends the authkey (**KIV_servK_authentic_ter**, omitted here). That event also binds peer A, so the conclusion does not need the existential quantification. Each version can be more convenient than the other to apply at times, but for simplicity we will make no distinction in the following.

Although the servticket has the same structure as the authticket, the former is sealed with the shared key belonging to an agent different from Tgs. The regularity lemma must be applied to investigate the authenticity of the servticket. We prove (Theorem 7.3.6) that Tgs indeed sent the servticket encrypted under a session key that originated with Kas.

Theorem 7.3.6 (KIV_servTicket_authentic). *If B is uncompromised and is not Tgs and evs is such that*

Crypt(shrK B){|Agent A, Agent B, Key $servK$, Number Ts|}
 ∈ parts(spies evs)

then, for some authK and Ta, evs contains

Says Tgs A (Crypt $authK$ {|Key $servK$, Agent B, Number Ts,
          Crypt(shrK B){|Agent A, Agent B, Key $servK$, Number Ts|}|}) *and*
Says Kas A (Crypt(shrK A){|Key $authK$, Agent Tgs, Number Ta,
          Crypt(shrK Tgs){|Agent A, Agent Tgs, Key $authK$, Number Ta|}|}).

7.3.4 Unicity

Since Kas issues fresh authkeys, the classical unicity theorem can be used to state that an authkey cannot appear in two different messages. Two other theorems rest on the Unique predicate (§4.4) to state that Kas only sends each authkey once, and that Tgs only sends each servkey once (**KIV_Kas_Unique** and **KIV_Tgs_Unique**, omitted here).

Both the authticket and the servticket have the same structure and contain a fresh session key, so we can establish a useful unicity result that applies to either ticket under the assumption that the session key is confidential (Theorem 7.3.7). Relaxing this assumption by the confidentiality argument (§7.3.5) makes the theorem applicable by either Tgs or B, and the corresponding goals available to them.

Theorem 7.3.7 (KIV_unique_CryptKey). *If evs is such that*

Key $K \notin$ analz(spies *evs*)

and

Crypt(shrK P){|Agent Q, Agent P, Key K, T|} $\in$ parts(spies *evs*)   *and*
Crypt(shrK P'){|Agent Q', Agent P', Key K, T'|} $\in$ parts(spies *evs*)

then

$$P = P' \quad and \quad Q = Q' \quad and \quad T = T'.$$

7.3.5 Confidentiality

Each authkey can be used to encrypt several servkeys because an agent, once authenticated, can request more than one service. Therefore, we expect that the compromise of an authkey may cascade to several servkeys. Conversely, servkeys are never used to encrypt any keys, so the compromise of a servkey should not affect other keys.

These observations can be proved to hold formally, providing three significant session key compromise theorems. They also become fundamental rewrite rules for the analz operator when proving the session key confidentiality theorems, the actual confidentiality goals of the protocol.

Session key compromise. Fragments of the proof script pertaining to the following guarantees can be found in Appendix A.2. The association of authkeys with servkeys in Kerberos IV resembles the association of session keys with nonces in the Yahalom protocol [135]. These relations can be formalised in a similar fashion. We declare the predicate

AKcryptSK : [key, key, event list]

and want it to hold for a trace featuring an event that associates an authkey with a servkey. It can be defined as

AKcryptSK *authK servK evs* ≜

  ∃ *A B Ts* |

    Says Tgs *A* (Crypt *authK* {|Key *servK*, Agent *B*, Number *Ts*,

                  Crypt(shrK *B*){|Agent *A*, Agent *B*,

                          Key *servK*, Number *Ts*|}|})

    ∈ set *evs*

It is clear that the association is established by Tgs. Several lemmas must be proved about the predicate that formalises it. Chiefly, no keys encrypt an authkey or a shared key (Lemma 7.3.4); a servkey does not encrypt any keys (Lemma 7.3.5); and only a single authkey encrypts a servkey (Lemma 7.3.6).

Lemma 7.3.4 (KIV_authKeys_are_not_AKcryptSK). *If evs is such that*

  K ∈ authKeys *evs* *or* K ∈ range shrK

then K ∈ symKeys *and, for any* K', *evs is such that*

  ¬AKcryptSK K' K *evs.*

Lemma 7.3.5 (KIV_not_authKeys_not_AKcryptSK). *If evs is such that*

  servK ∉ authKeys *evs* *and* *servK* ∉ range shrK

then, for any K, *evs is such that*

  ¬AKcryptSK *servK* K *evs.*

Lemma 7.3.6 (KIV_not_different_AKcryptSK). *If evs is such that*

  AKcryptSK *authK servK evs*

and K *is not authK, then servK* ∈ symKeys *and evs is such that*

  ¬AKcryptSK K *servK evs.*

If a session key K is not associated with a key K' by the predicate AKcryptSK, then the key K is never used to encrypt K', so the compromise of K should not increase the Spy's chances of discovering K'. Expressing this formally provides an essential lemma (Lemma 7.3.7), where K is generalised to a set KK. The logical disjunction has higher priority than the logical coimplication, as stated above (§1.3.3). As for notation, −(range shrK) indicates the set of all keys that are not shared keys, − being the set complement operator. While KK is a set of keys, Key`$KK$ yields the corresponding set of messages, ` being the image operator. The proof is rather long and complicated: simplification takes three quarters of the total computational time, and several case analyses are required afterwards. The assumptions of the theorem hold, for example, of an authkey K and a servkey K', provided that K has not been used to encrypt K'.

Lemma 7.3.7 (KIV_Key_analz_image_Key). *If* $K' \in$ symKeys *and*

$$KK \subseteq -(\text{range shrK})$$

and, for any $K \in KK$, *evs is such that*

$$\neg \text{AKcryptSK}\ K\ K'\ \text{evs}$$

then evs is such that

Key $K' \in$ analz(Key$`KK \cup$ (spies *evs*))   *if and only if*

$K' \in KK$ *or*   Key $K' \in$ analz(spies *evs*).

The first session key compromise theorem concerns authkeys and shared keys: they can be proved to be unaffected by the accidental loss of any session key (Theorem 7.3.8). The proof follows from applying Lemma 7.3.4 to Lemma 7.3.7. Recall that authkeys are particularly valuable secrets because of their long lifetime.

Theorem 7.3.8 (KIV_analz_insert_freshK1). *If evs is such that*

$$K \in \text{authKeys}\ \text{evs}\quad \text{or}\quad K \in \text{range shrK}$$

and $K' \notin$ range shrK, *then evs is such that*

Key $K \in$ analz($\{$Key $K'\} \cup$ (spies *evs*))   *if and only if*

$K = K'$ *or*   Key $K \in$ analz(spies *evs*).

It can also be proved that no cryptographic key is affected by the loss of a servkey (Theorem 7.3.9). Observe that no particular assumptions bind K, which is merely a symmetric key. The proof follows from the application of Lemma 7.3.5 to Lemma 7.3.7.

Theorem 7.3.9 (KIV_analz_insert_freshK2). *If evs is such that*

$$\text{servK} \notin \text{authKeys}\ \text{evs}\quad \text{and}\quad \text{servK} \notin \text{range shrK}$$

and $K \in$ symKeys, *then evs is such that*

Key $K \in$ analz($\{$Key *servK*$\} \cup$ (spies *evs*))   *if and only if*

$K = servK$ *or*   Key $K \in$ analz(spies *evs*).

Another theorem concerns the servkeys: given the authkey that is associated with a servkey, no other authkey would help the Spy discover the servkey (Theorem 7.3.10). The proof follows from applying Lemma 7.3.6 to Lemma 7.3.7. A variant of this theorem, where the definition of AKcryptSK is unfolded (**KIV_analz_insert_freshK3_bis**, omitted here), is more straightforward to use in practice.

Theorem 7.3.10 (KIV_analz_insert_freshK3). *If evs is such that*

$$\text{AKcryptSK}\ \text{authK}\ \text{servK}\ \text{evs}$$

and K *is not authK, and* $K \notin$ range shrK, *then evs is such that*

Key *servK* $\in$ analz($\{$Key $K\} \cup$ (spies *evs*))   *if and only if*

$servK = K$ *or*   Key *servK* $\in$ analz(spies *evs*).

Session key confidentiality. Fragments of the proof script pertaining to the following guarantees can be found in Appendix A.3. Session key confidentiality theorems express the confidentiality of the session keys. Following the general method, we first verify the property from the viewpoints of the Servers. We then refine the findings via the authenticity theorems to study conformity to the principle of goal availability.

If an authkey has not expired, then it cannot be lost by accident via the *OopsA* rule. Assuming that its peer is uncompromised assures that the key travels safely inside the second message of the protocol. Under these assumptions, the key confidentiality can be enforced from the viewpoint of Kas (Theorem 7.3.11) by frequent appeals to Theorems 7.3.8 and 7.3.10. Because Kas is reliable, the theorem does not need to insist that the encrypting key is A's shared key.

Theorem 7.3.11 (KIV_Confidentiality_Kas). *If A is uncompromised and evs contains*

Says Kas A (Crypt(shrK A){Key $authK$, Agent Tgs,
        Number Ta, $authTicket$})

and is such that ¬expiredAK Ta *evs, then evs is such that*

Key $authK$ ∉ analz(spies *evs*).

Refining this result by the authenticity Theorem 7.3.3 produces a confidentiality guarantee (**KIV_Confidentiality_Auth_A**, omitted here) that A can apply within her minimal trust. Hence, the protocol makes authkey confidentiality available to A in the stated threat model.

An analogous result enforces the confidentiality of the servkey from Tgs's viewpoint under the assumption that the key has not expired and that the authkey associated with it is confidential (Theorem 7.3.12). Frequent appeals to Theorems 7.3.8, 7.3.9 and 7.3.10 are necessary to the proof. The assumption of authkey confidentiality is clearly indispensable; otherwise, the fourth message of the protocol would falsify the conclusion of the theorem. That assumption can be relaxed by Theorem 7.3.11, producing a guarantee that is applicable by Tgs, since Kas and Tgs can inspect each other's functioning (**KIV_Confidentiality_Tgs_bis**, omitted here). These guarantees point out that Tgs's minimal trust includes the recipients of the servkey to be uncompromised; otherwise, they would reveal the key.

Theorem 7.3.12 (KIV_Confidentiality_Tgs). *If A and B are uncompromised and evs contains*

Says Tgs A (Crypt $authK$ {Key $servK$, Agent B, Number Ts, $servTicket$})

and is such that

Key $authK$ ∉ analz(spies *evs*)   *and*   ¬expiredSK Ts *evs*

then evs is such that

Key $servK \notin$ analz(spies evs).

There exists a guarantee of servkey confidentiality that A can apply upon reception of specific instances of the second and fourth messages (Theorem 7.3.13); hence that goal is available to A. It can be proved using existing authenticity and confidentiality theorems. The first step is the application of Theorem 7.3.3 to Theorem 7.3.11 to derive the authkey confidentiality. Then, Theorem 7.3.5 gives that the servkey originated with Tgs, and Theorem 7.3.12 concludes.

Theorem 7.3.13 (KIV_Confidentiality_Serv_A). *If A and B are uncompromised and evs is such that*

Crypt(shrK A){Key $authK$, Agent Tgs, Number Ta, $authTicket$}
   $\in$ parts(spies evs)   *and*

Crypt $authK$ {Key $servK$, Agent B, Number Ts, $servTicket$}
   $\in$ parts(spies evs)   *and*

$\neg$expiredAK Ta evs *and* $\neg$expiredSK Ts evs

then evs is such that

Key $servK \notin$ analz(spies evs).

Investigating the servkey confidentiality from B's viewpoint reveals a violation of the goal availability principle, which in turn provides the opportunity for novel insights (Theorem 7.3.14). The proof develops in a forward style. First, it elaborates on the given history of $servK$, deriving the confidentiality of $authK$ by Theorems 7.3.3 and 7.3.11. Then, Theorem 7.3.1 states that $authK$ is a session key (it does not belong to the range of shrK), which is necessary to apply Theorem 7.3.5 and derive the origin of $servK$. Certainly B is not a Kerberos Server, so he formally differs from Tgs; otherwise, Theorems 7.3.4 and 7.3.1 would derive that $servK$ is an authkey while Theorem 7.3.2 states that it is not. At this stage, Theorem 7.3.6 introduces another possible history of $servK$ but the unicity argument for Tgs unifies the two histories. An appeal to Theorem 7.3.12 concludes.

Theorem 7.3.14 (KIV_Confidentiality_B). *If A and B are uncompromised, B is not Tgs and evs is such that*

Crypt(shrK B){Agent A, Agent B, Key $servK$, Number Ts}
   $\in$ parts(spies evs)   *and*

Crypt $authK$ {Key $servK$, Agent B, Number Ts, $servTicket$}
   $\in$ parts(spies evs)   *and*

Crypt(shrK A){Key $authK$, Agent Tgs, Number Ta, $authTicket$}
   $\in$ parts(spies evs)   *and*

$\neg$expiredAK Ta evs *and* $\neg$expiredSK Ts evs

then evs is such that

Key $servK \notin$ analz(spies evs).

A closer look at the proof just described shows that, after Theorems 7.3.3, 7.3.11 and 7.3.5 are applied, the fact that $servTicket = \{\!| A, B, servK, Ts |\!\}_{Kb}$ could be derived by Theorem 7.3.2. Since parts is closed under message decomposition, the first assumption about trace evs is technically unnecessary. However, the present formulation highlights what B can or cannot verify.

In the real world, B can only witness the reception of the servticket, thus verifying the first and fifth assumptions about evs. He is certainly not able to verify the second, third and fourth because they pertain to the authentication and authorisation phases, which he does not participate in. With goal availability in mind, we wonder: are these assumptions necessary? They signify that the authkey that encrypts the servkey whose confidentiality is being studied has not expired. This is indispensable for the application of Theorems 7.3.3 and 7.3.11 and enforces the authkey confidentiality, which is itself indispensable for deriving the servkey confidentiality due to the form of the fourth message. The conclusion is that the protocol fails to make servkey confidentiality available to B in our realistic model. In consequence, Kerberos IV can be attacked as follows.

Attacking the protocol. Because of the assumptions that B cannot verify, Theorem 7.3.14 cannot be applied by B. The theorem reveals that, upon reception of the servticket, B must assume that the preceding phases have not compromised the authkey and consequently the servkey. But this clearly does not belong to his minimal trust. Therefore, the protocol can be attacked as follows.

1. The authkey belonging to A expires and the Spy gets hold of it, while A is killed and her owner logged out from the workstation.
2. The Spy extracts all servkeys associated with that authkey from messages she has previously intercepted. This is an attack on servkey confidentiality.
3. The Spy exploits each servkey not yet expired to spoof the corresponding instance of the fifth message (she only needs to update the timestamp of the authenticator) for some network service B.
4. The Spy terminates the service phase with B and gets access to its resources. This is an attack on authentication of A with B.

As often remarked above, an attack is subjected to a threat model. Our attacks succeed because the Spy can get hold of expired session keys (via the oops events). Clearly, an attack is as realistic as the underlying threat model. Session keys are often discarded in the real world when they expire, but remain in memory regions that are rarely overwritten. In consequence, our attacks become realistic. If a spoofed servkey has not yet expired, the Spy can obtain access for its remaining lifetime to the corresponding service, which in fact was granted to the initiator. The service believes to be communicating with the initiator, but the initiator does not exist anymore. Also, the initiator's owner is no longer connected and hence cannot register any irregularities.

Fixing the protocol. The attacks succeed because the servkeys may remain valid even after the authkey with which they are associated expires. Preventing this is a fix. It is sufficient to constrain Tgs's operation with a suitable temporal check so that it does not issue servkeys that would expire after the authkey associated with them. Our formal model can easily reflect the change by adding the temporal check

$$(\text{CT } evs) + \text{servKlife} \leq Ta + \text{authKlife} \tag{7.1}$$

to the preconditions of rule *KIV4*. Because *Ta* is the timestamp that marks the issue of the authkey, the check exactly prescribes that the expiry time of the servkey at the latest equals that of the authkey.

In the model for the updated protocol, Theorem 7.3.14 no longer needs the second, third and fourth assumptions, and so becomes applicable by *B* (**KIVu_Confidentiality_B**, omitted here). Therefore, the strengthened protocol makes servkey confidentiality available to *B*. Although shorter than the old proof, the new one requires some arithmetic to handle the temporal check. Moreover, *B* must be explicitly assumed not to be Tgs; otherwise, the servticket $\{A, B, servK, Ts\}_{Kb}$ could be misinterpreted as an authticket. Theorem 7.3.6, updated to enforce also condition 7.1, derives a history of *servK*. From ¬expiredSK *Ts evs* and condition 7.1, we derive ¬expiredAK *Ta evs*. Theorems 7.3.11 and then 7.3.12 conclude the proof. Observe that the unicity argument is not required because the reasoning only develops along a single history of *servK*.

7.3.6 Authentication

Authentication is a major goal of Kerberos IV, especially between the initiator and the network service. We pragmatically analyse what forms of this goal are achieved throughout the three phases of the protocol.

Authentication phase. Although the first phase is explicitly meant for authentication, the first message does not authenticate the initiator to Kas because it is a cleartext. The Spy might overload the Server with fake requests even if she were an outsider. The same can be observed of BAN Kerberos (§6.3), but these protocols are not meant to resist denial-of-service attacks. The authenticity Theorem 7.3.3 also authenticates Kas to the initiator *A*, providing her with an available guarantee of weak agreement with the first Kerberos Server.

Authorisation phase. The authenticator $\{A, T2\}_{authK}$ of the third message aims at authenticating *A* with Tgs. However, Tgs's merely receiving the authenticator does not enforce the goal: the Server must also receive the corresponding authticket $\{A, \text{Tgs}, authK, Ta\}_{Ktgs}$ to learn *authK* and be able to decipher the authenticator. If *authK* is confidential in this scenario, then *A* sent those certificates in an instance of the third message (Theorem 7.3.15). Relaxing the last assumption by the confidentiality Theorem 7.3.11 results in a guarantee applicable by Tgs. Observe that the regularity lemma guarantees the integrity of the authticket because Tgs is uncompromised, while the confidentiality of *authK* establishes integrity of the authenticator.

Theorem 7.3.15 (KIV_Tgs_authenticates_A). *If* *A* *is uncompromised and evs is such that*

Crypt *authK* {|Agent *A*, Number *T2*|} ∈ parts(spies *evs*)   *and*

Crypt(shrK Tgs) {|Agent *A*, Agent Tgs, Key *authK*, Number *Ta*|}
   ∈ parts(spies *evs*)   *and*

Key *authK* ∉ analz(spies *evs*)

then, for some B, evs contains

Says *A* Tgs {|Crypt(shrK Tgs){|Agent *A*, Agent Tgs, Key *authK*, Number *Ta*|},
   Crypt *authK* {|Agent *A*, Number *T2*|}, Agent *B*|}.

The theorem formalises weak agreement of *A* with Tgs but does not express that *A* is the true creator of the authenticator, which would give evidence to Tgs that *A* knows *authK* and was alive at time *T2*. Later (§8.4), we will prove formally that this stronger goal is met.

The proof of the last theorem relies on the integrity of the two certificates it mentions, which is guaranteed by the confidentiality of the encrypting keys. It requires substantial simplification to deal with the long inductive formula. One non-trivial subgoal arises from rule *KIV3*, as it introduces an event of the same form as that asserted by the theorem. If the two events contain different authkeys, the inductive formula terminates the proof; otherwise, an appeal to the unicity Theorem 7.3.7 is required. Another significant subgoal arises from rule *KIV5* because the form of the authenticator in the inductive formula matches that of the authenticator that this rule prescribes *A* to send to *B*. However, this implies that the authkey *authK* is being used in place of a servkey in the fifth message, which the protocol forbids. Precisely, we can track the origin of *authK* back to Kas by a lemma stating that *A* only invokes Tgs after Kas operated (**KIV_K3_imp_K2**, omitted here). Also, *authK* is tracked back to Tgs by Theorem 7.3.5. Finally, we apply Theorems 7.3.1 and 7.3.2 to derive the contradiction that *authK* both is and is not an authkey.

The authenticity Theorem 7.3.5 appears to be a relevant guarantee of what may seem weak agreement of Tgs with *A*. In studying whether that goal is available to *A*, we observe that the confidentiality assumption on the authkey can be relaxed by the corresponding argument (namely Theorem 7.3.11 refined by Theorem 7.3.3). However, the conclusion must quantify *A* existentially — hence the goal fails — because the premises do not bind her identity. Opening up the servticket would technically help because it would provide the necessary instance of A, although we would miss goal availability since A cannot inspect that certificate in practice. However, before concluding that the protocol fails to make the goal available to A, we must pragmatically verify if there are other ways to bind A using assumptions that A can check. One way is the reception of the second message, which Kas sends. Therefore, we can refine (variant **KIV_servK_authentic_ter** of) Theorem 7.3.5 by Theorem 7.3.3 and obtain a theorem confirming that weak agreement of Tgs with A is made available to A (Theorem 7.3.16).

Theorem 7.3.16 (KIV_A_authenticates_Tgs). *If A is uncompromised and evs is such that*

> Crypt(shrK A){Key $authK$, Agent Tgs, Number Ta, $authTicket$}
> $\in$ parts(spies evs)   *and*
> Crypt $authK$ {Key $servK$, Agent B, Number Ts, $servTicket$}
> $\in$ parts(spies evs)   *and*
> Key $authK$ $\notin$ analz(spies evs)

then evs contains

> Says Tgs A (Crypt $authK$ {Key $servK$, Agent B, Number Ts, $servTicket$}).

Service phase. The same method proves weak agreement of A with B (Theorem 7.3.17) via a lemma containing the full event whereby Tgs sends off the servticket (**KIV_Says_K5**, omitted here). The theorem relies on certificates of the same form as those of Theorem 7.3.15. However, assuming B not to be Tgs establishes that they contain a servticket. As expected, B must be assumed to be uncompromised in order for the servticket to be integral. Later (§8.4), the theorem will be strengthened to state formally that A knows $servK$ and was alive at time $T3$.

Theorem 7.3.17 (KIV_B_authenticates_A). *If A and B are uncompromised, B is not Tgs and evs is such that*

> Crypt $servK$ {Agent A, Number $T3$} $\in$ parts(spies evs)   *and*
> Crypt(shrK B){Agent A, Agent B, Key $servK$, Number Ts}
> $\in$ parts(spies evs)   *and*
> Key $servK$ $\notin$ analz(spies evs)

then evs contains

> Says A B {Crypt(shrK B){Agent A, Agent B, Key $servK$, Number Ts},
>   Crypt $servK$ {Agent A, Number $T3$}}.

Refining the assumption of servkey confidentiality of this theorem by Theorem 7.3.14 fails to make it applicable by B (**KIV_B_authenticates_A_r**, omitted here). By contrast, refining it by the corresponding theorem for the updated protocol would leave only assumptions that B can check within his minimal trust (**KIVu_B_authenticates_A_r**, omitted here). In conclusion, only the updated protocol makes the goal of weak agreement of A with B available to B. The original protocol in fact suffers the authentication attack described in the previous section.

Proving weak agreement of B with A (Theorem 7.3.18) is slightly more complicated because the last message of the protocol fails to state B's identity. Fortunately, this lack of explicitness can be overcome by additional checks on A's side. The theorem does not state formally that B knows $servK$ or when B was alive. These facts will be proved later (§8.4).

Theorem 7.3.18 (KIV_A_authenticates_B). *If A and B are uncompromised and evs is such that*

Crypt $servK$ (Number $T3$) $\in$ parts(spies evs)   *and*

Crypt $authK$ {|Key $servK$, Agent B, Number Ts, $servTicket$|}
   $\in$ parts(spies evs)   *and*

Crypt(shrK A){|Key $authK$, Agent Tgs, Number Ta, $authTicket$|}
   $\in$ parts(spies evs)

Key $authK$ $\notin$ analz(spies evs)   *and*   Key $servK$ $\notin$ analz(spies evs)

then evs contains

Says B A (Crypt $servK$ (Number $T3$)).

To understand this guarantee, let us recall that the last message of the protocol is a certificate sealed with a servkey. That key must be confidential for the certificate not to be a spoof. Upon reception of the certificate, A can derive that the servkey is meant for B by recalling the association established by an instance of the fourth message. This must be sealed with a confidential authkey meant for A in order for the association to be reliable, namely not invented by the Spy.

If we relax the confidentiality assumptions on both keys by the corresponding arguments seen above from A's viewpoint, we find out that the theorem only rests on assumptions that A can verify within her minimal trust (**KIV_A_authenticates_B_r**, omitted here). This lets us conclude that even the original protocol makes weak agreement of B with A available to A.

7.3.7 Key Distribution

The current potential of the Inductive Method is inadequate to reason about key distribution (§4.7), as observed during the analysis of BAN Kerberos (§6.5.7). In the next chapter, we will introduce the necessary extensions to verify this goal (§8.4).

8. Modelling Agents' Knowledge of Messages

Knowledge of messages for agents other than the Spy is defined via possession of the messages. Two formalisations in the Inductive Method are given and compared. They serve to treat a strong form of authentication (non-injective agreement) and key distribution.

Reasoning formally about security protocols invites confusion between belief and knowledge. Abstracting from the context may help clarify the difference. If Alice *knows* something, then she has sufficient evidence that it is true. If she *believes* something, then she does not necessarily have evidence about it. The BAN logic [58] only captures the notion of belief, though this is often misinterpreted as knowledge. The predicate P believes ϕ signifies that "P would be entitled to believe X" and that "P may act as though X is true" [58, p. 236]. Subsequent research extends the logic with a proper concept of knowledge: Bleeker and Meertens [52] introduce the predicate P rightly_believes ϕ, which holds when P believes ϕ and ϕ in fact holds.

As demonstrated in the previous chapters, the Inductive Method already provides a meta-formalisation of agents' beliefs and knowledge by means of theorems. Those stated from an agent's viewpoint express the agent's knowledge if the agent can verify all their assumptions. Conversely, if the agent cannot obtain evidence of the truth values of some assumptions, then the theorems only express the agent's beliefs. Strictly speaking, even when some assumptions belong to the minimal trust (Chapter 5), the beliefs cannot become knowledge.

The present chapter provides a formal definition of the agents' knowledge of messages and message components, rather than knowledge that certain events occurred. This is crucial to verify the main goal of most of the protocols already formalised: key distribution. Each of the peers of a session key should receive guarantees that the other peer knows the same key. Moreover, the definition prepares the Inductive Method for analysing new hierarchies of protocols. For example, non-repudiation protocols [170, 171] aim at non-repudiation of reception, which requires proving agents' knowledge of specific components upon their reception (Chapter 13). E-commerce protocols [37] will not authorise delivery of goods until the merchant obtains assurances that his bank knows the components that correspond to the exact payment.

Protocols for group key agreement aim at distributing a key to all agents of the group, providing each agent with evidence that the goal is met [18, 136].

We express knowledge of a component via possession of the component [24]. This notion can be formalised via the ability to actively use the component, which is in turn verifiable by inspecting the history of the network (§8.1). Alternatively, reception of the component also expresses its possession but requires introducing the corresponding event (§8.2). This also improves the readability of the analyses. The outcomes of both approaches are presented in detail on BAN Kerberos (§8.3) and on Kerberos IV (§8.4), producing stronger guarantees of authentication and novel guarantees of key distribution. The chapter continues with a comparison of the two approaches (§8.5), and finally discusses the outcomes of using timestamps or nonces on the same protocol design (§8.6), an issue that requires reasoning on agents' knowledge. Theorems about various protocol versions and various protocol models are discussed; hence, adherence to our theorem naming conventions (§1.3.2) is particularly useful.

8.1 Agents' Knowledge via Trace Inspection

The agent who creates a message certainly knows all components of the message, including the cryptographic key that possibly seals the message. Unfortunately, a Says event is inadequate to express creation (as repeatedly observed above, in §§4.6, 6.5.6, and 7.3.6) because it may occur when an agent is merely forwarding an unintelligible message.

However, if we require that a message X never appeared, even within a larger message, before the event Says $A\,B\,X$ occurs, then A is the true creator of X. This can be established by inspecting the history that precedes the event. Since each history is recorded by a trace, enforcing the property simply requires a trace inspection. We declare the predicate

Issues : [agent, agent, msg, event list]

so that A Issues B with X on evs holds when A creates message X for B in a trace evs. Events may occur more than once and recent events are added at the head of traces. Therefore, detecting the first occurrence through time of an event ev in a trace evs requires scanning evs in reverse, namely from tail to head. Traces are lists, and the theory file List.thy of the Isabelle distribution [33, 34] provides numerous functions for reasoning about lists. The unary rev reverses a list, while the binary takeWhile takes a predicate and a list, scans the list and returns all elements of the list until these verify the predicate. Other known functions are useful to define Issues as

A Issues B with X on $evs \triangleq$

$\quad \exists Y.$ Says $A\,B\,Y \in$ set evs and $X \in$ parts$\{Y\}$ and

$\quad X \notin$ parts(spies (takeWhile($\lambda x.\,x \neq$ Says $A\,B\,Y$)(rev evs)))

It can be seen that the predicate requires that A send X, possibly inside a larger message, and that X never appear in the traffic preceding such an event. Hence, the predicate can be used not only to convey agents' knowledge but also to formalise agent authentication.

8.1.1 Basic Lemmas

A few technical lemmas are necessary to reason about reversed lists. All of them are provable by induction on the relevant trace. They do not depend on the specific protocol under analysis as they are merely trace properties. For example, those from theory file KerberosIV.thy (Figure 3.1) are quoted here, but *evs* is a generic trace, not necessarily of the protocol model.

The traffic in a trace whose oldest element is a Says event amounts to the traffic on the subtrace without the event plus the message introduced by that event (Lemma 8.1.1). Recall that @ is the Isabelle symbol for the list concatenation operator.

Lemma 8.1.1 (KIV_spies_Says_rev).
Trace evs is such that spies$(evs$ @ $[$Says $A\,B\,X]) = \{X\} \cup ($spies $evs)$.

Together with an analogous law concerning the Notes event (omitted here), Lemma 8.1.1 serves to establish an obvious result formally: the traffic on a reversed trace is the same as that on the original trace (Lemma 8.1.2).

Lemma 8.1.2 (KIV_spies_evs_rev).
Trace evs is such that spies $evs =$ spies$($rev $evs)$.

Another technically important result states that the traffic on a subtrace obtained via the takeWhile function is a subset of the traffic on the whole trace (Lemma 8.1.3).

Lemma 8.1.3 (KIV_spies_takeWhile).
Trace evs is such that spies$($takeWhile $P\,evs) \subseteq ($spies $evs)$.

The monotonicity law for parts can be resolved with Lemma 8.1.3, obtaining a result (Lemma 8.1.4) that will be very useful below.

Lemma 8.1.4 (KIV_parts_spies_takeWhile_mono).
Trace evs is such that parts(spies$($takeWhile $P\,evs)) \subseteq$ parts$(($spies $evs))$.

8.1.2 Proving Knowledge

A general method can be devised to prove significant guarantees in terms of the Issues predicate.

Let us suppose that a message X is a component of a larger message Y, and that the event Says $A\,B\,Y$ occurs in a trace *evs*. If A is the true creator of X, then (and only then) can we prove that A Issues B with X on *evs* holds. Some assumptions are necessary to assure that the Spy cannot forge X before

A actually creates it: if X is a concatenated message, then we must assume that the Spy cannot analyse its components from the traffic or synthesise them; if X is a ciphertext, then we must assume confidentiality of the key that seals it. Also, A must be assumed not to be the Spy, so she acts legally according to the protocol. Further assumptions may be required on A and B (§§8.3.1 and 8.4.1) depending on the protocol. The proof develops through the following method.

- Simplify the main subgoal by the definition of Issues.
- Isolate the first two conjuncts of the definition of Issues by proceeding in a backward style (resolving first by the introduction rules for existential quantification, and then by conjunction).
- Prove the first conjunct, Says $A\,B\,Y$, by assumption, and the second conjunct, $X \in$ parts$\{Y\}$, by symbolic evaluation of the parts operator.
- Apply structural induction over the protocol model to verify that all steps of the protocol definition preserve the following property: the occurrence of the event Says $A\,B\,Y$ in a trace implies that Y never appears in the trace before that event.
- Simplify all subgoals.
- Prove the subgoal corresponding to the protocol step where the event Says $A\,B\,Y$ takes place by applying Lemmas 8.1.2 and 8.1.3, a few other trivial ones, and a lemma introducing the event Says $A\,B\,Y$ on the available assumptions (§§8.3.1 and 8.4.1).

Although such a theorem has limited importance in itself, since it merely says that an agent knows the components of the messages he sends, it is particularly useful to refine other theorems that enforce the event Says $A\,B\,Y$. For example, using this method we have investigated the goals of non-injective agreement and key distribution on all protocols analysed so far, as demonstrated on two Kerberos versions in the rest of this chapter.

8.2 Agents' Knowledge via Message Reception

The extensions to the Inductive Method mentioned in this section were released with the 1999 distribution of Isabelle [158]. They can be found in file Event.thy (Figure 3.1).

A primitive Gets can be introduced to model message reception [22]. The Isabelle datatype of events (§3.7) must be extended as

> datatype event $\triangleq$ Says agent agent msg
>
> Notes agent msg
>
> Gets agent msg

In the real world, a message can be received only if it was previously sent. This *reception invariant* can be easily enforced by the protocol model (§8.2.2).

Technically speaking, the Notes event could be replaced by the Gets event imagining that, when an agent notes down a message, it is as if the agent received it from the network. However, this would compromise the reception invariant. Keeping the two events separate allows reasoning that turns out to be more readable, more faithful to reality and, ultimately, simpler. For example, thanks to the reception invariant, the messages that are received do not need to enrich the set of components used in a trace. The definition of used (§3.10) must be enriched with the following rule.

3. All messages received in a trace do not directly extend those that are used on that trace.

   used$((\mathsf{Gets}\,A\,X)\,\#\,evs) \triangleq$ used evs

8.2.1 From Spy's Knowledge to Agents' Knowledge

The major outcome of introducing message reception is a realistic formalisation of agents' knowledge. The rudimentary version available initially within the Inductive Method [132] was soon specialised to express knowledge of a single agent, the Spy, by means of the function spies (§3.9). The previous chapters have demonstrated that this function allows reasoning about confidentiality but not about key distribution. We generalise spies to a binary function

knows : [agent, event list] $\longrightarrow$ msg set

It is meant to capture any agent's knowledge: the first parameter represents the agent whose knowledge is being defined. The entire range of the function initState (§3.9) now becomes relevant to the rest of the treatment. The definition of knows generalises that of spies in the base case and the two inductive steps corresponding to the existing events. A third inductive step becomes necessary to account for the Gets event.

0. An agent knows his initial state.

   knows $A\,[] \triangleq$ initState A

1. An agent knows what he sends to anyone in a trace; in particular, the Spy also knows all messages ever sent on it.

   knows $A\,((\mathsf{Says}\,A'\,B\,X)\,\#\,evs) \triangleq$
   $$\begin{cases} \{X\} \cup \text{knows } A \; evs & \text{if } A = A' \text{ or } A = \mathsf{Spy} \\ \text{knows } A \; evs & \text{otherwise} \end{cases}$$

2. An agent knows what he notes in a trace; the Spy also knows the compromised agents' notes.

knows A $((\text{Notes } A' X) \# evs)$ $\triangleq$

$$\begin{cases} \{X\} \cup \text{knows } A \ evs & \text{if } A = A' \text{ or} \\ & (A = \text{Spy and } A' \in \text{bad}) \\ \text{knows } A \ evs & \text{otherwise} \end{cases}$$

3. An agent, except the Spy, knows what he receives in a trace. The Spy's knowledge must not be extended with any of the received messages since the Spy already knows them by case 1 and by the reception invariant.

knows A $((\text{Gets } A' X) \# evs)$ $\triangleq$

$$\begin{cases} \{X\} \cup \text{knows } A \ evs & \text{if } A = A' \text{ and } A \text{ is not the Spy} \\ \text{knows } A \ evs & \text{otherwise} \end{cases}$$

The initial hints to agents' knowledge [132, §4.5] suggested that an agent A can *see* a message X when X has been sent by someone to A. However, sending X off to A does not assure that X is delivered to A. The introduction of the reception event allows a faithful treatment of the matter through the last step of the definition of knows.

Recall that the function analz is applied to a set of messages and recursively extracts all components of concatenated messages and bodies of messages encrypted under keys that are known. If, in the real world, an agent A knows a message X, the protocol model contains a trace evs such that either $X \in (\text{knows } A \ evs)$ if A did not need any decryption to get hold of X, or $X \in \text{analz}(\text{knows } A \ evs)$ if A had to retrieve X from within a larger message of the set knows A evs. Therefore, also agents other than the Spy may now need to apply the function analz in some circumstances.

8.2.2 Updating the Existing Models

Let prot be a formal protocol model. The reception invariant (that a message can be received only if it was sent) can be enforced by adding an inductive rule to the definition of the protocol. The rule, which is called *Reception* (Figure 8.1), allows the extension of a trace of prot containing an event Says $A B X$ with the event Gets $B X$.

Reception :
⟦ evsR ∈ prot; Says A B X ∈ set evsR ⟧ ⟹ Gets B X # evsR ∈ prot

---

Fig. 8.1. Rule template for message reception

Observe that, since rules are never forced to fire on any trace, no Gets event can be forced to take place. Therefore, no guarantee can be established that a message that was sent will be ever received by the intended recipient, as is realistic in a setting where the Spy can prevent message delivery. Furthermore, each rule can fire more than once so that the same message can be received more than once. Rules can fire in the wrong order so that the order in which messages are sent is not necessarily preserved upon their reception. Observe also that agents discard the sender label upon reception because it is not reliable on an insecure network.

Fake:
```
⟦ evsF ∈ bankerb_gets; X ∈ synth (analz (knows Spy evsF)) ⟧
⟹ Says Spy B X # evsF ∈ bankerb_gets
```

Reception:
```
⟦ evsR ∈ bankerb_gets; Says A B X ∈ set evsR ⟧
⟹ Gets B X # evsR ∈ bankerb_gets
```

BK2:
```
⟦ evs2 ∈ bankerb_gets; Key K ∉ used evs2;
  Gets Server {|Agent A, Agent B|} ∈ set evs2 ⟧
⟹ Says Server A (Crypt (shrK A) {|Number (CT evs2), Agent B, Key K,
                  Crypt (shrK B) {|Number (CT evs2), Agent A, Key K|}|})
      # evs2 ∈ bankerb_gets
```

BK3:
```
⟦ evs3 ∈ bankerb_gets;
  Says A Server {|Agent A, Agent B|} ∈ set evs3;
  Gets A (Crypt (shrK A) {|Number Ts, Agent B, Key K, Ticket|})
    ∈ set evs3;
  ¬ expiredS Ts evs3 ⟧
⟹ Says A B {|Ticket, Crypt K {|Agent A, Number (CT evs3)|}|}
      # evs3 ∈ bankerb_gets
```

Fig. 8.2. Inductive model, updated with message reception, of BAN Kerberos: fragment

The protocol model needs further updates. If the event Says $P\,Q\,X$ appears in the premises of a rule and P does not appear in the conclusions, then the event can be replaced by Gets $Q\,X$. In the old model, Q acted when some other (undefined) agent had sent him X. However, no agent but the Spy can monitor events performed by other agents, so the condition was not directly verifiable by Q until X was received. Thus, the new model expresses Q's behaviour more accurately. Figure 8.2 shows a fragment of the updated model for the BAN Kerberos, called bankerb_gets. It is derived from that in Figure 6.2 (§6.4) and can be obtained with file Kerberos_BAN_Gets.thy (Figure 3.1). It can be seen that the *Fake* rule features knows Spy rather than spies — all occurrences of spies should be updated similarly. Observe also the Gets preconditions in rules *BK2* and *BK3*.

8.2.3 Basic Lemmas

For example, let us consider a generic trace *evs* of the protocol model bankerb_gets just seen, though the following treatment holds for all protocols where the Gets event is used as suggested. The reception invariant can be easily proved by induction (Lemma 8.2.1).

Lemma 8.2.1 (BKg_Gets_imp_Says).
If evs contains Gets $B\,X$, *then, for some A, evs also contains* Says $A\,B\,X$.

Applying this lemma and a basic one stating that the Spy knows the messages that have been sent, we can prove that she also knows all messages that have been received (Lemma 8.2.2).

Lemma 8.2.2 (BKg_Gets_imp_knows_Spy).
If evs contains Gets $B\,X$, *then evs is such that* $X \in ($knows Spy *evs*$)$.

Applying this lemma when B is the Spy, we derive that the Spy knows the messages she receives. The last case of the definition of knows allows this result to be generalised: any agent knows what he receives (Lemma 8.2.3).

Lemma 8.2.3 (BKg_Gets_imp_knows).
If evs contains Gets $B\,X$, *then evs is such that* $X \in ($knows B *evs*$)$.

Resolving Lemma 8.2.2 with the lemma $H \subseteq$ parts H, we obtain that messages that are received by anyone appear in the traffic (Lemma 8.2.4).

Lemma 8.2.4 (BKg_Gets_imp_knows_Spy_parts).
If evs contains Gets $B\,X$, *then evs is such that* $X \in$ parts(knows Spy *evs*$)$.

Resolving Lemma 8.2.3 with the lemma $H \subseteq$ analz H, we obtain that messages that an agent receives really are accessible to that agent (Lemma 8.2.5). This is important because analz is idempotent: analz(analz H) = analz H. So, if a message is accessible, it can be accessed with finite efforts.

Lemma 8.2.5 (BKg_Gets_imp_knows_analz).
If evs contains Gets $B\,X$, *then evs is such that* $X \in$ analz(knows B *evs*$)$.

8.2.4 Updating the Existing Theorems

Lemma 8.2.4 introduces significant modifications to some of the theorems already proved so far. In consequence, some theorems become available (in the sense of goal availability) to the agents. For example, the unicity Theorem 6.5.4 for BAN Kerberos can be made available to agents who are uncompromised (Theorem 8.2.1). Here, *evs* still is a generic trace of bankerb_gets. The proof rests on a double application of Lemma 8.2.4; a double application of the authenticity Theorem 6.5.2 then introduces the necessary assumptions to apply the unicity Theorem 6.5.4.

Theorem 8.2.1 (BKg_unique_session_keys_Gets). *If A is uncompromised and evs contains*

Gets A (Crypt(shrK A){|Number Tk, Agent B, Key Kab, $Ticket$|}) *and*

Gets A (Crypt(shrK A){|Number Tk', Agent B', Key Kab, $Ticket'$|})

then

$$Tk = Tk' \quad and \quad B = B' \quad and \quad Ticket = Ticket'.$$

Therefore, should an agent who is uncompromised receive the same session key within two different messages, she could suspect that something has gone wrong (something that is outside our model). Arguably, the Unique predicate cannot be proved to hold on any reception event, for any message can be received more than once (§8.2.2).

Other theorems become more readable and faithful to reality thanks to the reception event. For example, authentication of A to B for Kerberos IV can be established upon B's reception of a suitable message, as we can see below (Theorem 8.4.7). It certainly is clearer than the analogous theorem without the reception event, which was given in the previous chapter (Theorem 7.3.17).

8.2.5 Proving Knowledge

The crucial fact about agents' knowledge is Lemma 8.2.5: any agent knows what he receives (and its components). Hence, when designing guarantees for an agent A, we can inform her of B's knowledge of all components of X by proving that an event Gets B X took place. However, proving that the event occurred depends on the protocol under analysis. The rest of this chapter shows how to prove the goals of authentication and key distribution for BAN Kerberos (§8.3.2) and for Kerberos IV (§8.4.2).

8.3 Revisiting the Guarantees on BAN Kerberos

We have tested both approaches to agents' knowledge on all existing protocol analyses. This section presents the outcomes on the verification of BAN Kerberos. As agents' knowledge is modelled, the most significant benefits concern key distribution and, indirectly, a stronger form of authentication than that discussed in Chapter 6: non-injective agreement on the session key. Minor outcomes have already been mentioned (§8.2.4).

In this section, for the sake of readability, when the assumption that a session key is confidential is necessary, it is left unrelaxed (though it can be relaxed by the appropriate confidentiality argument seen above).

8.3.1 Using Trace Inspection

Here, *evs* is a generic trace of `bankerberos`. The general method (§8.1.2) can be applied to prove that if an uncompromised agent sends an instance of the third message of BAN Kerberos, then that agent indeed created the message (Theorem 8.3.1).

Theorem 8.3.1 (BK_A_Issues_B). *If A and B are uncompromised and evs contains*

$$\text{Says}\,A\,B\,\{\!|\,Ticket, \text{Crypt}\,Kab\{\!|\text{Agent}\,A, \text{Number}\,Ta|\!\}|\!\}$$

and is such that

$$\text{Key}\,Kab \notin \text{analz}(\text{spies}\;evs)$$

then evs is such that

$$A\;\text{Issues}\;B\;\text{with}\,(\text{Crypt}\,Kab\{\!|\text{Agent}\,A, \text{Number}\,Ta|\!\})\;\text{on}\;evs.$$

```
[ A ∉ bad; B ∉ bad; evs3 ∈ bankerb_gets;
  Says S A (Crypt (shrK A) {|Number Tk, Agent B, Key Kab, Ticket|})
    ∈ set evs3;
  Says A Server {|Agent A, Agent B|} ∈ set evs3; ¬ expiredK Tk evs3;
  Key Kab ∉ analz (spies evs3);
  Says A B {|Ticket, Crypt Kab {|Agent A, Number (CT evs3)|}|} ∉ set evs3;
  Crypt Kab {|Agent A, Number (CT evs3)|}
    ∈ parts
        (spies
          (takeWhile
            (λz. z ≠ Says A B {|Ticket,
                                Crypt Kab {|Agent A, Number (CT evs3)|}|})
              (rev evs3))) ]
⟹ False
```

Fig. 8.3. Proving an Issues property for BAN Kerberos

As anticipated by the general method, the most difficult subgoal to prove arises from the case formalising the third step of the protocol, which is quoted in Figure 8.3. It can be seen that the last assumption of the subgoal can be resolved with Lemma 8.1.4, producing

$$\text{Crypt}\,Kab\{\!|\text{Agent}\,A, \text{Number}(\text{CT}\;evs3)|\!\} \in \text{parts}(\text{spies}\,(\text{rev}\,evs3))$$

which, resolved with Lemma 8.1.2, means that the authenticator appears in the traffic. At this stage, the proof would terminate if we could apply the authentication Theorem 6.5.9. Precisely, its variant **BK_B_authenticates_A**, where the confidentiality assumption on the session key has not yet been relaxed by Theorem 6.5.8, seems useful. But not all of its assumptions are yet available, as it is missing the fact that the ticket in its intelligible form is in the traffic, that is

Crypt(shrK B){Number Tk, Agent A, Key Kab} $\in$ parts(spies $evs3$)

So, we try to derive this fact. From the fourth assumption of the subgoal, it follows that $evs3$ is such that

Crypt(shrK A){Number Tk, Agent B, Key Kab, $Ticket$} $\in$ parts(spies $evs3$)

By Theorem 6.5.2, the fourth assumption in the figure also holds when the agent S is the Server. Then, Theorem 6.5.1 derives the form of the ticket. Because the Server sent it, it is in the traffic, so the assumption that we were missing has now been found to hold. Hence, the proof proceeds as anticipated above and the last subgoal terminates as indicated.

The mentioned variant of the authentication Theorem 6.5.9 can be resolved with the first assumption of Theorem 8.3.1. This produces a guarantee for B conveying A's knowledge of Kab and non-injective agreement of A with B on Kab (Theorem 8.3.2), because the Issues predicate hides and enriches the corresponding Says event. It becomes available to B if we relax the confidentiality assumption. The theorem assumes A to be uncompromised, so she acts legally. This implies that she was alive at time Ta. As we point out below, the theorems presented here also confirm the key distribution goal; hence their names.

Theorem 8.3.2 (BK_B_authenticates&keydist_to_A). *If A and B are uncompromised and evs is such that*

Crypt(shrK B){Number Tk, Agent A, Key Kab} $\in$ parts(spies evs)   *and*
Crypt Kab{Agent A, Number Ta} $\in$ parts(spies evs)   *and*
Key $Kab \notin$ analz(spies evs)

then evs is such that

A Issues B with (Crypt Kab{Agent A, Number Ta}) on *evs.*

The same procedure relies on the authentication Theorem 6.5.10 to prove an analogous guarantee for A (Theorem 8.3.3). Precisely, in this case we need variant **BK_A_authenticates_B** of Theorem 6.5.10, where the confidentiality assumption on the session key has not yet been relaxed by Theorem 6.5.7. The following theorem informs A that B knows Kab and establishes non-injective agreement of B with A on Kab.

Theorem 8.3.3 (BK_A_authenticates&keydist_to_B). *If A and B are uncompromised and evs is such that*

Crypt(shrK A){Number Tk, Agent B, Key Kab, $Ticket$}
   $\in$ parts(spies evs)   *and*
Crypt Kab(Number Ta) $\in$ parts(spies evs)   *and*
Key $Kab \notin$ analz(spies evs)

then evs is such that

B Issues A with $(\mathsf{Crypt}\,Kab(\mathsf{Number}\,Ta))$ on evs.

Unfortunately, the protocol only requires B to reply with an incremented Ta, so A only understands that B was alive after Ta. In other words, Ta is being used as a nonce. However, if B inserted the current time in place of Ta, then A would be informed of the exact instant when B was alive, which is a desirable outcome from the use of timestamps.

Let us summarise. Theorems 8.3.2 and 8.3.3 formally prove that BAN Kerberos also achieves the goal of key distribution, which completes the protocol analysis presented above (Chapter 6). Their confidentiality assumptions can be easily relaxed using the appropriate confidentiality argument, so we can conclude that the protocol makes key distribution available to its peers in our threat model featuring the temporal modelling of accidents. The same conclusion holds for mutual non-injective agreement on the session key if we consider these theorems, which express the necessary knowledge of the key, along with the authentication Theorems 6.5.9 and 6.5.10. In the next section, we shall study the same goals using message reception.

8.3.2 Using Message Reception

A similar reasoning can be conducted using the other approach to agents' knowledge, which is based on message reception. Recall that our aim is to prove that each of the peers received the session key on assumptions that the other peer can verify (§8.2.5). For the sake of demonstration, we consider the BAN Kerberos protocol and a generic trace evs of its formal model that makes use of message reception, `bankerb_gets`.

Let us suppose that an agent B who is uncompromised receives an instance of the third message of the protocol that quotes an agent A who is not the Spy. Assuming that this session key is confidential, it can be concluded that the session key must be known to A and that it was A who sent that message to B (Theorem 8.3.4).

Theorem 8.3.4 (BKg_B_authenticates&keydist_to_A). *If A and B are uncompromised and evs contains*

> Gets B $\{\!\!\{\mathsf{Crypt}(\mathsf{shrK}\,B)\{\!\!\{\mathsf{Number}\,Tk, \mathsf{Agent}\,A, \mathsf{Key}\,Kab\}\!\!\},$
> $\qquad \mathsf{Crypt}\,Kab\{\!\!\{\mathsf{Agent}\,A, \mathsf{Number}\,Ta\}\!\!\}\}\!\!\}$

and is such that

> Key $Kab \notin \mathsf{analz}(\mathsf{knows}\,\mathsf{Spy}\,evs)$

then evs contains

> Says $A\,B$ $\{\!\!\{\mathsf{Crypt}(\mathsf{shrK}\,B)\{\!\!\{\mathsf{Number}\,Tk, \mathsf{Agent}\,A, \mathsf{Key}\,Kab\}\!\!\},$
> $\qquad \mathsf{Crypt}\,Kab\{\!\!\{\mathsf{Agent}\,A, \mathsf{Number}\,Ta\}\!\!\}\}\!\!\}$

and is such that

Key $Kab \in$ analz(knows A evs).

The proof is as follows. Take Theorem 6.5.9, which authenticates A with B, precisely its mentioned variant where the key confidentiality assumption is yet unrelaxed. It can be updated according to the guidelines given above (§8.2.4), so that its two main assumptions become the single assumption that a suitable Gets event occurred. Theorem 6.5.9 concludes that A sent the right instance of the third message. Then, having that A is uncompromised, she certainly is not the Spy, and in consequence she must have received an instance of the second message containing Kab (**BKg_BK3_imp_Gets**, omitted here). Therefore, she can extract the session key by Lemma 8.2.5, because she knows her own shared key.

Not only does Theorem 8.3.4 establish key distribution to A but also non-injective agreement of A with B on Kab. The confidentiality assumption on the session key can be relaxed by Theorem 6.5.8, so that both goals become available to B.

The same method can be followed to prove analogous guarantees for A (Theorem 8.3.5).

Theorem 8.3.5 (BKg_A_authenticates&keydist_to_B). *If A and B are uncompromised and evs contains*

Gets A (Crypt Kab(Number Ta)) *and*

Gets A (Crypt(shrK A){Number Tk, Agent B, Key Kab, $Ticket$})

and is such that

Key $Kab \notin$ analz(knows Spy evs)

then evs contains

Says B A (Crypt Kab(Number Ta))

and is such that

Key $Kab \in$ analz(knows B evs).

The proof is similar to the previous one. Take the variant of the authentication Theorem 6.5.10 where the confidentiality assumption yet is unrelaxed, and update it according to the usual guidelines (§8.2.4). It enforces that B sent the correct instance of the fourth message. We can further infer that he received an instance of the third message because he is uncompromised (**BKg_BK4_imp_Gets**, omitted here). Therefore, he can learn the session key by an appeal to Lemma 8.2.5, because he knows his own shared key.

Not only does Theorem 8.3.5 establish key distribution to B but it also establishes non-injective agreement of B with A on Kab. The confidentiality assumption on the session key can be relaxed by Theorem 6.5.7, so that both goals become available to A.

Theorems 8.3.4 and 8.3.5 confirm the findings of the previous section, where a different modelling of agents' knowledge was used: BAN Kerberos makes key distribution and mutual non-injective agreement on the session key available to its peers in our threat model.

8.4 Revisiting the Guarantees on Kerberos IV

Both approaches to modelling agents' knowledge scale up to proving non-injective agreement and key distribution on Kerberos IV. The proofs are, as expected, longer than those for BAN Kerberos because distinguishing between the two kinds of session keys and tickets often requires case analyses. Also here, for the sake of readability, the key confidentiality assumptions are left unrelaxed.

8.4.1 Using Trace Inspection

Here, *evs* is a generic trace of kerbIV, the protocol model that comes with the file KerberosIV.thy (Figure 3.1). As previously mentioned, the authenticity Theorem 7.3.3 establishes weak agreement of Kas with A. The general method (§8.1.2) may be used to establish that, if Kas sends an instance of the second message, then Kas is indeed issuing it, namely the message never appeared before (**KIV_Kas_Issues_A**, omitted here). Resolving the assumption of this result by Theorem 7.3.3 provides A with evidence that Kas knows the authkey mentioned in the message (**KIV_A_authenticates&keydist_to_Kas**, omitted here). This establishes non-injective agreement of Kas with A on the authkey. As we shall see, these theorems also confirm various forms of key distribution; hence their names.

If A is an uncompromised agent sending an instance of the third message, then we can prove that A is the creator of the authenticator that the message contains (**KIV_A_Issues_Tgs**, omitted here). Observe that, while Kas certainly is not the Spy thanks to the injections created by the Isabelle datatype of messages (§3.6), the generic agent A must be explicitly assumed to be uncompromised in order for the result to hold. Combining this with the authentication Theorem 7.3.15, we obtain a guarantee of non-injective agreement of A with Tgs on the authkey (Theorem 8.4.1). Relaxing the confidentiality assumption on the authkey by Theorem 7.3.11, the guarantee can be applied by Tgs (within its minimal trust that A is uncompromised). The theorem also informs Tgs that A was alive at time $T2$.

Theorem 8.4.1 (KIV_Tgs_authenticates&keydist_to_A). *If A is uncompromised and evs is such that*

Crypt $authK$ {|Agent A, Number $T2$|} $\in$ parts(spies *evs*)   *and*

Crypt(shrK Tgs){|Agent A, Agent Tgs, Key $authK$, Number Ta|}
   $\in$ parts(spies *evs*)   *and*

Key $authK$ $\notin$ analz(spies *evs*)

then evs is such that

A Issues Tgs with (Crypt *authK* {|Agent *A*, Number *T2*|}) on *evs*.

Like Kas, also Tgs acts legally. So, we can prove that if Tgs sends an instance of the fourth message, then the message is new (**KIV_Tgs_Issues_A**, omitted here). This result can be combined with Theorem 7.3.16, which expressed weak agreement of Tgs with *A*, arriving at a guarantee of non-injective agreement of Tgs with *A* on the authkey and the servkey (Theorem 8.4.2). The same guarantee also tells *A* that Tgs was alive at time *Ts*.

Theorem 8.4.2 (KIV_A_authenticates&keydist_to_Tgs). *If A is uncompromised and evs is such that*

Crypt(shrK *A*){|Key *authK*, Agent Tgs, Number *Ta*, *authTicket*|}
    ∈ parts(spies *evs*) *and*
Crypt *authK* {|Key *servK*, Agent *B*, Number *Ts*, *servTicket*|}
    ∈ parts(spies *evs*) *and*
Key *authK* ∉ analz(spies *evs*)

then evs is such that

Tgs Issues *A* with
    (Crypt *authK* {|Key *servK*, Agent *B*, Number *Ts*, *servTicket*|}) on *evs*.

We can also prove that an uncompromised agent who sends an instance of the fifth message does create the authenticator included in the message (**KIV_A_Issues_B**, omitted here). Its assumptions can be relaxed by Theorem 7.3.17, deriving a guarantee of non-injective agreement of *A* with *B* on the servkey (Theorem 8.4.3), which also tells *B* that *A* was alive at time *T3*.

Theorem 8.4.3 (KIV_B_authenticates&keydist_to_A). *If A and B are uncompromised, B is not* Tgs *and evs is such that*

Crypt *servK* {|Agent *A*, Number *T3*|} ∈ parts(spies *evs*) *and*
Crypt(shrK *B*){|Agent *A*, Agent *B*, Key *servK*, Number *Ts*|}
    ∈ parts(spies *evs*) *and*
Key *servK* ∉ analz(spies *evs*)

then evs is such that

A Issues *B* with (Crypt *servK* {|Agent *A*, Number *T3*|}) on *evs*.

Finally, if an uncompromised agent *B* sends an instance of the last message sealed with a confidential servkey, then the message can be proved to be new (**KIV_B_Issues_A**, omitted here). This result can be used to refine Theorem 7.3.18, thereby expressing non-injective agreement of *B* with *A* on the servkey (Theorem 8.4.4). Also, *B* does not cheat because he is assumed to be uncompromised. Therefore, as observed on BAN Kerberos, *A* learns that

B was alive after $T3$ but would get more precise information if B replaced $T3$ with a fresh timestamp. The protocol merely prescribes B to increment $T3$, an imprecise feature that we chose not to model (§7.2.4).

Theorem 8.4.4 (KIV_A_authenticates&keydist_to_B). *If A and B are uncompromised, B is not* Tgs *and evs is such that*

Crypt $servK$ (Number $T3$) $\in$ parts(spies evs)   *and*

Crypt $authK$ {Key $servK$, Agent B, Number Ts, $servTicket$}
   $\in$ parts(spies evs)   *and*

Crypt(shrK A){Key $authK$, Agent Tgs, Number Ta, $authTicket$}
   $\in$ parts(spies evs)   *and*

Key $authK$ $\notin$ analz(spies evs)   *and*   Key $servK$ $\notin$ analz(spies evs)

then evs is such that

B Issues A with (Crypt $servK$ (Number $T3$)) on evs.

The theorems presented here can be also interpreted to confirm key distribution. Theorems 8.4.3 and 8.4.4 signify that the protocol guarantees key distribution of a servkey to its peers. Theorems 8.4.1 and 8.4.2 signify that it also meets key distribution of an authkey to the protocol initiator and to Tgs. Relaxing the confidentiality assumptions by the appropriate formal arguments, we can conclude that these key distribution goals are available to the mentioned agents. In the next section, we shall see how to study the same goals by message reception.

8.4.2 Using Message Reception

Here, evs is a generic trace of the protocol model kerbIV_gets, which comes with the file KerberosIV_Gets.thy (Figure 3.1). The entire hierarchy of theorems presented below rests on a single philosophy: the authenticity (§7.3.3) or authentication (§7.3.6) theorems updated by suitable reception events (§8.2.4) provide guarantees of weak agreement. Using the definition of knows, we prove that specific agents have knowledge of specific session keys. Combining these guarantees, we derive non-injective agreement on the session keys, and also key distribution.

If an agent who is uncompromised receives the instance of the second message that is sealed with her shared key, an appeal to Lemma 8.2.4 derives the necessary assumptions to apply Theorem 7.3.3. The resulting theorem, stating that Kas sent the message received by that agent, establishes weak agreement of Kas with A. Since Kas knows all shared keys, it can extract and learn the authkey contained in the message. The combination of these results (**KIVg_A_authenticates&keydist_to_Kas**, omitted here) guarantees non-injective agreement of Kas with A on the authkey. Formally, it also expresses key distribution of the authkey to Kas, a guarantee that is typically taken for granted — in fact, the obvious technical lemma that Kas knows all agents' shared keys had never been proved before (**KIVg_shrK_in_knows_Server**, omitted here).

The updated authentication Theorem 7.3.15 states that, upon reception by Tgs of an instance of the third message that includes a confidential authkey, the agent mentioned by the authenticator sent the message. Let A be such agent. If A is uncompromised, induction proves that A sends the third message only upon reception of the second (**KIVg_K3_imp_Gets**, omitted here), from which she can extract the authkey. Hence, Tgs can be assured that A knows the authkey (Theorem 8.4.5). This theorem assures non-injective agreement of A with Tgs on the authkey, and also assures key distribution of the authkey to A.

Theorem 8.4.5 (KIVg_Tgs_authenticates&keydist_to_A). *If A is uncompromised and evs contains*

Gets Tgs {|Crypt(shrK Tgs){|Agent A, Agent Tgs, Key $authK$, Number Ta|},
    Crypt $authK$ {|Agent A, Number $T2$|}, Agent B|}

and is such that

Key $authK \notin$ analz(knows Spy evs)

then, for some B, evs contains

Says A Tgs {|Crypt(shrK Tgs){|Agent A, Agent Tgs, Key $authK$, Number Ta|},
    Crypt $authK$ {|Agent A, Number $T2$|}, Agent B|}

and is such that

Key $authK \in$ analz(knows A evs).

The existential form of the conclusion derives from application of Theorem 7.3.15. The identity of the intended recipient of the third message cannot be confirmed because the message is concatenated; hence, the Spy can alter that identity while the message is in the network. Therefore, Tgs cannot have the identity of A's intended recipient confirmed.

The authentication Theorem 7.3.16 can now be proved assuming the protocol initiator's reception of the suitable messages. We already observed (§7.3.6) that, if A is the initiator, then the guarantee conveys weak agreement of Tgs with A because it states that Tgs sent the message to A. Because Tgs only acts legally, it must have received the suitable instance of the third message (**KIVg_K4_imp_Gets**, omitted here), learning the authkey from the authticket by Lemma 8.2.5. Also, by definition of knows, Tgs knows the servkey that it has associated with the authkey. Agent A can be therefore informed that Tgs knows both session keys (Theorem 8.4.6). We have thus established non-injective agreement of Tgs with A on the authkey and the servkey, and also key distribution of both keys to Tgs.

Theorem 8.4.6 (KIVg_A_authenticates&keydist_to_Tgs). *If A is uncompromised and evs contains*

Gets A (Crypt(shrK A){Key $authK$, Agent Tgs, Number Ta, $authTicket$})

and

Gets A (Crypt $authK$ {Key $servK$, Agent B, Number Ts, $servTicket$})

and is such that

Key $authK \notin$ analz(knows Spy evs)

then evs contains

Says Tgs A (Crypt $authK$ {Key $servK$, Agent B, Number Ts, $servTicket$})

and is such that

Key $authK \in$ analz(knows Tgs evs)   *and*
Key $servK \in$ analz(knows Tgs evs).

Theorem 7.3.17, expressing weak agreement of A with B, can be updated as usual to rely on an appropriate Gets event. The theorem concludes that A sent the right instance of the fifth message. Continuing on this premise, it follows that, if A is uncompromised, she must have received a suitable instance of the fourth message (**KIVg_K5_imp_Gets**, discussed below), and must have extracted the servkey as we will see. Therefore, we have developed a guarantee of non-injective agreement of A with B on the servkey, and also of key distribution of the servkey to A (Theorem 8.4.7).

Theorem 8.4.7 (KIVg_B_authenticates&keydist_to_A). *If A and B are uncompromised, B is not* Tgs *and evs contains*

Gets B {Crypt(shrK B){Agent A, Agent B, Key $servK$, Number Ts},
    Crypt $servK$ {Agent A, Number $T3$}}

and is such that

Key $servK \notin$ analz(knows Spy evs)

then evs contains

Says A B {Crypt(shrK B){Agent A, Agent B, Key $servK$, Number Ts},
    Crypt $servK$ {Agent A, Number $T3$}}

and is such that

Key $servK \in$ analz(knows A evs).

This theorem is slightly more complicated to prove than others of the same form. Observe that the fifth message cannot bind the authkey because the receiver B is not meant to see any such keys. In consequence, a lemma like **KIVg_K5_imp_Gets** necessarily must quantify that key existentially. We need to infer that the initiator A knows the very authkey of which the theorem merely conveys existence. A quick inspection of rule *KIV5* of the protocol model (Figure 7.6) shows that A's sending the fifth message also implies that she sent the corresponding instance of the third (Lemma 8.4.1).

Lemma 8.4.1 (KIVg_K5_imp_Gets). *If* A *is uncompromised and evs contains*

Says A B {| $serv\,Ticket$, Crypt $servK$ {|Agent A, Number $T3$|}|}

then, for some authK, Ts, authK and T2, evs contains

Gets A (Crypt $authK$ {|Key $servK$, Agent B, Number Ts, $serv\,Ticket$|}) *and*
Says A Tgs {| $auth\,Ticket$, Crypt $authK$ {|Agent A, Number $T2$|}, Agent B|}.

The first event of the conclusion confirms that we only need to express A's knowledge of the existentially quantified authkey to terminate our reasoning, because that would allow A to extract the servkey. The second event of the conclusion can be shown by an appropriate lemma (**KIVg_K3_imp_Gets**, omitted here) to imply that A received that very authkey within an instance of the second message. Because this message is encrypted using A's shared key, she is able to extract the session key, appealing to Lemma 8.2.5.

Finally, the authentication Theorem 7.3.18 can be updated in terms of reception of the messages it mentions. Following the usual method, it can be formally proved that an agent who is uncompromised sends only the last message of the protocol upon reception of the last but one (**KIVg_K6_imp_Gets**, omitted here). He can therefore extract the servkey from the received servticket. This method informs the initiator A that the servkey she has received is also known to the responder B (Theorem 8.4.8). This guarantees non-injective agreement of B with A on the servkey, and also key distribution of the servkey to B.

Theorem 8.4.8 (KIVg_A_authenticates&keydist_to_B). *If* A *and* B *are uncompromised and evs contains*

Gets A {|Crypt $authK$ {|Key $servK$, Agent B, Number Ts, $serv\,Ticket$|},
      Crypt $servK$ (Number $T3$)|} *and*
Gets A (Crypt(shrK A){|Key $authK$, Agent Tgs, Number Ta, $auth\,Ticket$|})

and is such that

Key $authK$ $\notin$ analz(knows Spy evs) *and* Key $servK$ $\notin$ analz(knows Spy evs)

then evs contains

Says B A (Crypt $servK$ (Number $T3$)))

and is such that

Key $servK$ $\in$ analz(knows B evs).

8.5 Comparing the Two Approaches

One aim of formalising agents' knowledge was to investigate the goals of non-injective agreement and key distribution. We have verified them on all classical protocols analysed so far (for example, Needham-Schroeder, Yahalom, Otway-Rees and Woo-Lam) using both our approaches to knowledge (§§8.1 and 8.2).

The analysis of BAN Kerberos supports the claim that the two approaches are equivalent. Theorems 8.3.2 and 8.3.3, obtained by trace inspection, appear to convey the same guarantees as Theorems 8.3.4 and 8.3.5 respectively, which are obtained by message reception. Also Kerberos IV confirms the equivalence: Theorems 8.4.1 to 8.4.4 appear to be equivalent to Theorems 8.4.5 to 8.4.8 respectively.

According to our principle of goal availability (Chapter 5), no theorem has practical relevance unless its assumptions are verifiable. The guarantees proved by either approach become practically applicable at the same time, namely upon reception of the suitable messages. Nevertheless, one approach inherently lacks this level of formal detail.

Since the approach based on trace inspection cannot rely on a formalisation of the instant of reception, it prescribes scanning a trace to pinpoint a suitable sending event, and hence may seem technically more complicated. However, the other approach requires deriving the very reception event whereby an agent learns the message component under study, typically a session key (see the mentioned theorems of form **BKg_X_imp_Gets** and **KIVg_Y_imp_Gets**). This is not always easy during our proofs. For example, a message sealed with the session key is often available, but reception of such a message may only confuse our aim because there obviously exists no rule to derive knowledge of the encrypting key from knowledge of a ciphertext. So, the two approaches seem equally complicated.

The guarantees obtained by trace inspection on protocols based on time-stamps also add a temporal requisite to the goal of authentication: establishing that an agent who is uncompromised creates a message containing the current time as a timestamp also expresses *when* the agent was alive. Therefore, this approach will be used to compare the temporal requisites that timestamps or nonces add to a protocol design (§8.6), because the other approach clearly is not suitable to the purpose.

Both versions of Kerberos require both peers to use a session key to create new messages. On the contrary, let us consider a protocol that delivers a session key to a peer without requiring the peer to use it. Since the approach based on trace inspection expresses knowledge of a message (and its components) via the ability to create the message, it cannot be used to prove any significant properties in this case. By contrast, the approach based on message reception seems more appropriate to use in this case: it can express the peer's knowledge of the session key upon its reception, thus allowing investigation of key distribution and non-injective agreement.

Recall our strong definition of key distribution (§4.7). We shall see (§8.5.1) that if the delivery of the session key is not the last step of the protocol, then certain forms of key distribution can be met depending entirely on the protocol design. Conversely, we can conclude that when the delivery of a session key to an agent is the last step of the protocol, key distribution to that agent is not available to the peer because none of the protocol events can be used to prove that the session key was received.

8.5.1 On Otway-Rees and Otway-Rees-Bella

The Otway-Rees protocol [58, p. 244] offers another comparison. The responder B obtains the session key from the Server and forwards it to the initiator A in the last message of the protocol (Figure 8.4).

$$
\begin{aligned}
&1. \quad A \;\rightarrow\; B \;\;:\;\; M, A, B, \{Na, M, A, B\}_{Ka} \\
&2. \quad B \;\rightarrow\; S \;\;:\;\; M, A, B, \{Na, M, A, B,\}_{Ka}, \{Nb, M, A, B\}_{Kb} \\
&3. \quad S \;\rightarrow\; B \;\;:\;\; M, \{Na, Kab\}_{Ka}, \{Nb, Kab\}_{Kb} \\
&4. \quad B \;\rightarrow\; A \;\;:\;\; M, \{Na, Kab\}_{Ka}
\end{aligned}
$$

Fig. 8.4. Otway-Rees protocol

The protocol is easy to understand, its third message being the most important. It sees the Server send B two certificates respectively sealed with A and B's long-term keys, each containing a copy of the session key. So, B learns the session key from decrypting the second certificate, and forwards the rest to A. Upon reception of the last message, A learns the session key and binds it to B having seen Na returned, which she initially associated with B. Similarly, B binds the session key to A having seen Nb returned, which he associated with A in the second message. This might have been an insecure choice because A's identity arrived in the clear from the first message, but the Server confirms the association by matching the contents of the two cypher-texts received in the second message.

Because the protocol delivers the session key to A in the last message, key distribution of the session key to A certainly is not available to B. The other half of the goal, key distribution to B, seems more interesting to study. However, we cannot use trace inspection here because B is simply forwarding a certificate, containing the session key, which already is in the traffic. We can take the approach based on message reception instead.

The analysis of the protocol by the BAN logic led to the following conclusion: "it is interesting to note that this protocol does not make use of Kab as an encryption key, so neither agent can know whether the key is known to the other" [58, p. 247]. More precisely, the BAN logic cannot derive the usual conclusion that

$$
A \models B \models A \xleftrightarrow{Kab} B
$$

We observe that the Spy can intercept the third message, building the fourth on her own and sending it off to A (Figure 8.5). This qualifies as an attack on key distribution to A.

$$3. \quad \mathsf{S} \;\to\; B \;:\; M, \{\!|Na, Kab|\!\}_{Ka}, \{\!|Nb, Kab|\!\}_{Kb} \quad \text{(intercepted)}$$
$$4'. \quad \mathsf{Spy} \;\to\; A \;:\; M, \{\!|Na, Kab|\!\}_{Ka}$$

Fig. 8.5. Key distribution attack on Otway-Rees

Thanks to our experience with the verification of the various Kerberos protocols, we decide to update the third message of Otway-Rees so that encryption under B's shared key includes the other encryption. The rest is left unaltered. We address the resulting protocol as Otway-Rees-Bella (Figure 8.6). Like Otway-Rees, the updated protocol makes no use of the session key as an encryption key, though it makes key distribution to A available to B. In fact, the BAN logic fails to capture knowledge of the messages that are received, as does our approach based on trace inspection. However, using message reception, we discover that the reason Otway-Rees misses the goal is entirely due to message design.

$$3. \quad \mathsf{S} \;\to\; B \;:\; M, \{\!|\{\!|Na, Kab|\!\}_{Ka}, Nb, Kab|\!\}_{Kb}$$

Fig. 8.6. Otway-Rees-Bella protocol: fragment

Our claims on the Otway-Rees-Bella protocol can be confirmed formally. Below, *evs* is a generic trace of the updated protocol model, `orb`, which comes with the file `OtwayReesBella.thy` (Figure 3.1). The new form of the third message allows us to track B's participation. The Spy cannot substitute him because she cannot decrypt the message. More formally, A's certificate is kept confidential until B extracts it and sends the last message (Theorem 8.5.1). Observe that assuming the Says event, which A can verify in practice, is indispensable to binding the identity of the responder B.

Theorem 8.5.1 (ORB_analz_hard). *If A and B are uncompromised and evs contains*

Says A B {|Nonce M, Agent A, Agent B,
        Crypt(shrK A){|Nonce Na, Nonce M, Agent A, Agent B|}|}

and is such that

Crypt(shrK A){|Nonce Na, Key Kab|} $\in$ analz(knows Spy *evs*)

then evs contains

Says B A {|Nonce M, Crypt(shrK A){|Nonce Na, Key Kab|}|}.

The same result cannot be proved for the original Otway-Rees, where the assumption holds also before B acts. Also, the result fails to hold if we replace analz with parts because the certificate appears as a component of the traffic even before B sends the last message. The proof is long, the subgoal arising from the last protocol step being particularly complicated. That subgoal requires simplifying the assumption

Crypt(shrK A) {|Agent B, Nonce M, Nonce Na, Key Kab|}
    $\in$ analz({Key K} $\cup$ (knows Spy $evs4$))

Its symbolic evaluation is not trivial. A rewriting rule can be proved to inform the simplifier that no session key is used to encrypt a certificate in the protocol (Lemma 8.5.1).

Lemma 8.5.1 (ORB_analz_insert_freshCryptK). *If evs is such that*

Key $K \notin$ analz(knows Spy evs)

and

$K' \notin$ range shrK

then evs is such that

Crypt $K X \in$ analz({Key K'} $\cup$ (knows Spy evs)) *if and only if*
Crypt $K X \in$ analz(knows Spy evs).

The proof can be developed through the conventional method for the session key compromise theorem (§4.5). However, it requires several subsidiary results concerning the form of the message sent by the Server, and of that received by B, which apparently originated with the Server.

Another important portion of reasoning asserts that an agent who is uncompromised sends the last message of the protocol only upon reception of an integral, suitable instance of the last but one (Theorem 8.5.2).

Theorem 8.5.2 (ORB_OR4_imp_Gets). *If B is uncompromised and evs contains*

Says B A {|Nonce M, Crypt(shrK A) {|Nonce Na, Key Kab|}|}

then, for some Nb, evs contains

Gets B {|Nonce M, Crypt(shrK B) {|Crypt(shrK A) {|Nonce Na, Key Kab|},
                    Nonce Nb, Key Kab|}|}.

We now have all fragments of A's reasoning. Upon reception of the last message of the protocol, A concludes that the certificate is available to the Spy by an analogous of Lemma 8.2.2 proved for this protocol, then by $H \subseteq$ analz H and finally by message decomposition. This means that the second assumption of Theorem 8.5.1 holds. By an appeal to that theorem, A derives that B indeed participated in the protocol by sending the last message. Theorem 8.5.2 can now be applied, confirming that B received an intelligible message quoting the same session key received by A. Certainly, B can decrypt a message sealed with his own shared key, and extract the session key. The resulting guarantee confirms to A that the session key she receives is also known to her peer B (Theorem 8.5.3).

Theorem 8.5.3 (ORB_A_keydist_to_B). *If A and B are uncompromised and evs contains*

Says A B {|Nonce M, Agent A, Agent B,
            Crypt(shrK A){|Nonce Na, Nonce M, Agent A, Agent B|}|}

and

Gets A {|Nonce M, Crypt(shrK A){|Nonce Na, Key Kab|}|}

then evs is such that

Key Kab $\in$ analz(knows B evs).

The final theorem lets us conclude that our updated Otway-Rees protocol makes the goal of key distribution to B available to A. It only differs from the original protocol in the form of the message issued by the Server. Nevertheless, as with Otway-Rees, our protocol *does not make use of Kab as an encryption key* but does inform A that B knows the session key. Hence, our protocol falsifies the BAN logic claim reported above.

We have learnt that prescribing no use of a session key after its delivery must not be taken as generally undermining availability of key distribution.

8.5.2 On Public-key Protocols

Testing our approaches to agents' knowledge on public-key protocol unveils limitations of each one. For example, let us consider the first step of the Needham-Schroeder protocol and suppose that it takes place during the history modelled by the trace *evs*. Then, *evs* contains the event

Says A B (Crypt(pubK B){|Agent A, Nonce Na|})

whereby agent A issues a nonce Na and sends it to B inside a certificate sealed with B's public key. Since A does not know B's private key, which is necessary to decrypt the certificate, we cannot establish that Na belongs to analz(knows A evs). However, A in fact knows Na because she just created it. Trace inspection regains its attraction in this case because it can be used to prove that A issues the entire certificate and therefore knows its components. Nonetheless, B will also know Na upon reception of the cipher, but this again requires reasoning with message reception. Upon event

Gets B (Crypt(pubK B){|Agent A, Nonce Na|})

we can formally prove that Na belongs to analz(knows B evs) because B knows the private key to decrypt the certificate.

Therefore, a combination of the two approaches, which can be easily implemented, will yield the best results when analysing public-key protocols.

8.6 Timestamps Versus Nonces on the Same Design

Timestamps form a linear order whereas nonces merely are random numbers; hence, it can be expected that the former may convey stronger temporal guarantees on a same protocol design. First, we provide an informal account (§8.6.1) and then we support it formally (§8.6.2).

8.6.1 Informal Account

With timestamps, any agent can check the freshness of a message containing a timestamp at any point, even without having yet participated in the protocol. Furthermore, a timestamp typically informs of *the exact time when* the message it accompanies was created, something that nonces cannot accomplish. However, three assumptions must be met. Firstly, agents must know the specific lifetime for the timestamp, though lifetimes are in general not secret. Secondly, and very importantly, the agent who inserted the timestamp must be trusted to have acted honestly; otherwise, he could insert a later timestamp than the current one to try and give the message a longer validity. Thirdly and most importantly, all agents must run a synchronisation protocol to synchronise their clocks. It could be argued that the threats to the synchronisation protocol might offset any gains obtained from using timestamps in the security protocol, but this matter is outside the focus of our research.

Nonces can be used to express message freshness. An agent is required to first participate in the protocol by sending a fresh nonce at a time instant with respect to which he requires a guarantee of freshness of a later message. An example comes from the shared-key Needham-Schroeder protocol, which was presented above (Figure 2.6). Its first two steps are recalled here (Figure 8.7).

$$
\begin{array}{llllll}
1. & A & \rightarrow & S & : & A, B, Na \\
2. & S & \rightarrow & A & : & \{\!|Na, B, Kab, \{\!|Kab, A|\!\}_{Kb}|\!\}_{Ka}
\end{array}
$$

Fig. 8.7. Shared-key Needham-Schroeder protocol: fragment

It can be seen that A issues the fresh nonce Na with the first message and sees it returned with the second, in which the Server is required to copy it. Because the nonce is random, upon reception of the second message, she concludes that the message was issued *some time after* she issued the first one. She does not learn when exactly the second message was created, but can decide to discard it should it be received too late with respect to her issue of the nonce. Clearly, the more the nonce is truly random, the more this reasoning is reliable.

8.6.2 Formal Account

Our aim here is to develop some formal argument to support or undermine the informal discussion anticipated above. Finding a suitable benchmark was not too difficult. A fair comparison of the outcomes of timestamps to those of nonces requires an analogous protocol layout. BAN Kerberos and the shared-key Needham-Schroeder protocol seem appropriate, as the former may be viewed exactly as the latter *"modified with the addition of timestamps"* [121]. We will compare the temporal requisites of the goals of message authenticity and agent authentication as they are achieved by the two protocols. The other goals are time independent. Our experiments formally confirm that time-stamps provide stronger temporal requisites to trusted agents on the same protocol design, provided that the agents' clocks are kept synchronised [39].

Recall the Needham-Schroeder protocol from Figure 2.6 (§2.2.4) and the BAN Kerberos protocol from Figure 6.1 (§6.3). The authenticity arguments for the two protocols are easy to compare. For brevity, we omit most of the formal syntax here as we have variously demonstrated it above. The complete treatment comes with the files `Kerberos_BAN.thy` and `NS_Shared.thy` (Figure 3.1).

The authenticity argument for BAN Kerberos (Theorem 6.5.2) assures an uncompromised protocol initiator A that the message

$$\{ Tk, B, Kab, Ticket \}_{Ka}$$

originated with the Server. Since the Server operates reliably, the guarantee informs A that the message and the session key it contains were created at time Tk. In the same setting, the message

$$\{ Na, B, Kab, Ticket \}_{Ka}$$

of the Needham-Schroeder protocol can be proved to have originated with the Server [33, 34]. Due to Na's being A's nonce, A is assured that the message is more recent than the instant when she issued Na. This temporal requisite also applies to the session key Kab. Though not identical, the two theorems appear to convey equivalent guarantees.

The two protocols provide different temporal requisites to the respective responders. An uncompromised protocol responder B can be assured that the ticket

$$\{ Tk, A, Kab \}_{Kb}$$

of the BAN Kerberos protocol originated with the Server (Theorem 6.5.3). By the presence of the timestamp, B derives that the ticket and the session key were created at time Tk, because the Server can be relied upon. No such guarantee is available to B when running Needham-Schroeder, because the ticket has the form

$$\{ Kab, A \}_{Kb}$$

Even if we prove that the ticket originated with the Server, B obtains no information on how recent Kab is. This is due to the fact that B did not participate earlier in the protocol and hence could not insert his own nonce.

With regard to the goal of authentication, Theorem 8.3.2 can be also invoked to formally assure B that A, who must be assumed uncompromised so that he acts legally, was alive at time Ta. An analogous guarantee we have proved for Needham-Schroeder conveys a similar requisite (Theorem 8.6.1). Here, evs is a generic trace of the formal protocol model **ns_shared**. The double concatenation of Nb formalises the last message of the protocol. Precisely, B is informed that A was alive after B created his nonce Nb.

Theorem 8.6.1 (NSS_B_authenticates&keydist_to_A). *If A and B are uncompromised, evs is such that*

Crypt(shrK B){|Key Kab, Agent A|} $\in$ parts(spies evs)   *and*
Crypt Kab {|Nonce Nb, Nonce Nb|} $\in$ parts(spies evs)   *and*
Key Kab $\notin$ analz(spies evs)

then evs is such that

A Issues B with (Crypt Kab {|Nonce Nb, Nonce Nb|}) on evs.

Likewise, Theorem 8.3.3 can be also invoked to formally assure A that B, who must be assumed uncompromised, was alive after time Ta. Precisely, B does not update Ta and hence is using the timestamp as a nonce. Had he updated it, A would have been informed of the exact time when he was alive. An analogous guarantee we have proved for Needham-Schroeder (Theorem 8.6.2) must be interpreted with care.

Theorem 8.6.2 (NSS_A_authenticates&keydist_to_B). *If A and B are uncompromised, evs is such that*

Crypt(shrK A){|Nonce Na, Agent B, Key Kab, $Ticket$|}
    $\in$ parts(spies evs)   *and*
Crypt Kab(Nonce Nb) $\in$ parts(spies evs)   *and*
Key Kab $\notin$ analz(spies evs)

then evs is such that

B Issues A with (Crypt Kab(Nonce Nb)) on evs.

It appears that A is simply informed that B creates a message containing the nonce Nb. Since Nb was issued by B, A cannot obtain any temporal requisite about B's operation directly from the use of nonces. However, she indirectly gets a similar guarantee. Because the session key is confidential, an assumption that can be relaxed by the appropriate confidentiality argument, A knows that B learnt the session key because she sent it to him. The conclusion of the theorem confirms that B is using the session key (to issue a certificate). In consequence, A learns that B was alive at some later time after she sent him the session key. We emphasise that this reasoning must not be ascribed to the use of nonces.

Observe that our approach to agents' knowledge based on trace inspection turns out to be particularly suitable to conduct the reasoning presented here, while the other approach is not much help in this case. The two theorems presented about the Needham-Schroeder protocol can be proved using the same method discussed above for theorems of the same form. Indeed, the two theorems also qualify as authentication (precisely, non-injective agreement on the session key) and key distribution guarantees, which can be made available to the respective peers by relaxing the confidentiality assumptions. It also follows that all Issues theorems seen above can be variously interpreted to express freshness of the involved messages or, likewise, to add a temporal requisite to authentication.

Our findings formally support the informal arguments sketched above: nonces appear to convey weaker temporal guarantees than timestamps if used on the same protocol design. However, if the design can be enriched to accommodate both peers' insertion of nonces, then the guarantees become comparable, as confirmed here for the protocol initiator.

Such protocols making a more complicated use of nonces do exist. One important example is the Yahalom protocol [58, p. 257], whose reliance on nonces is not straightforward. The protocol uses symmetric cryptography and essentially aims at distributing a session key to its peers with the help of a trusted Server (Figure 8.8).

$$
\begin{array}{llll}
1. & A & \rightarrow & B & : & A, Na \\
2. & B & \rightarrow & S & : & B, \{\!|A, Na, Nb|\!\}_{Kb} \\
3. & S & \rightarrow & A & : & \{\!|B, Kab, Na, Nb|\!\}_{Ka}, \{\!|A, Kab|\!\}_{Kb} \\
4. & A & \rightarrow & B & : & \{\!|A, Kab|\!\}_{Kb}, \{\!|Nb|\!\}_{Kab}
\end{array}
$$

Fig. 8.8. Yahalom protocol

We observe that Paulson's formal guarantees of agent authentication about this protocol provide significant temporal requisites to both peers [135]. They can be found with file Yahalom.thy of the Isabelle repository [33, 34].

Precisely, a theorem holds (**Y_YM3_auth_B_to_A**, omitted here), confirming that when A receives the third message, B has already sent the corresponding instance of the second. This confirms to A that B sent the second message after she issued the nonce Na with the first message, because B has used Na.

An analogous guarantee exists for B (**Y_YM4_imp_A_said_YM3**, omitted here). It confirms that, if B sent the second message and receives the last message made with a confidential session key, then it was A who sent him the last message. It also tells B that A was alive after he invented the nonce Nb with the second message, because A has used Na.

9. Verifying Another Deployed Protocol

Version V of Kerberos makes a lighter use of encryption than Version IV. Formal analysis confirms that both versions essentially achieve the same goals. An important byproduct is the development of new proof methods and new unicity guarantees.

Version V [101] is the latest deployed Kerberos protocol. It is based on the same layout as that of the previous versions but poses different requirements on the environment in which it is to run. It also is fundamentally different in terms of design because of a different use of encryption. For these reasons, many consider it an entirely new protocol and, therefore, worthy of formal analysis.

The changes in terms of working requirements are published [102]. We will briefly review them in this chapter, pointing out those that are relevant to formal verification. As we have often pointed out throughout this book, building a realistic formal model is fundamental to developing realistic proofs of correctness. It is therefore important to scrutinise which features of the new protocol version can and/or must be modelled formally. The major differences that the new version introduces are in terms of design: the tickets are no longer doubly encrypted, as we detail in the next section.

Kerberos notoriously is designed to operate on small-scale networks such as LANs. Each LAN can be seen as a Kerberos *realm*. Each realm has a Kas-Tgs pair, and it is possible to use authorisation credentials granted by Tgs on the local realm to access another Tgs on a remote realm. Kerberos V supports such inter-realm operations better than the earlier version by allowing for a more efficient — hierarchical — management and transmission of the Servers' long-term secrets between realms. However, such management and transmission strictly speaking are not part of the actual Kerberos protocol, as they are performed out of band. In consequence, they will not be part of our analysis.

Our model Kerberos Servers are not formally bound to any specific realm, and our inductive protocol model allows for various instances of each protocol message to be repeated indefinitely over each trace. It may however be interesting to explicitly model an entire array of Kerberos Servers, indefinitely long, perhaps using techniques that Paulson developed for the BULL recursive protocol [131]. These ideas will be subjects of future research if their importance is confirmed.

This chapter details the Kerberos V protocol (§9.1), its inductive model (§9.2) and its proved guarantees (§9.3). The presentation skims over those details that are common with Kerberos IV, which we have already seen in Chapter 7. Wherever appropriate, comparative treatment between the two versions is provided.

9.1 The Kerberos V Protocol

We begin the presentation of the protocol by outlining its new working requirements.

Kerberos V is independent of the encryption algorithm. This is implemented as a separate software module that the administrator can appropriately insert, while Kerberos IV only used DES, which may suffer export limitations. Hence, each ciphertext of the new version carries an identifier specifying which encryption algorithm was used to create it. A similar evolution concerns the reliance on a specific underlying network protocol, which was only the Internet Protocol in the old version.

Additional changes of working requirements pertain to the ticket lifetimes. While the lifetime used to be implemented as an 8-bit number of five minute units, resulting in a maximum of approximately 21 hours, Kerberos V carries a starting date and an expiry date for each ticket, a feature that can account for lifetimes of any length. Other changes are added support for credential forwarding from one host to another, and for renewing tickets. Also the naming system is more structured in Kerberos V.

Interoperability increases in the new version thanks to the standardisation [94] of the encoding philosophy for multi-byte values in protocol messages. This prevents the chance of a sender and a receiver encoding the value in their own respective native order, which would require a conversion.

The protocol is in Figure 9.1. It can be seen that all steps are inherited unvaried from the previous version, except for steps 2 and 4, where the tickets are not doubly encrypted. In particular, Kas sends the authticket as the second component of the second message, while the first component is the same as the entire old message though it does not include the authticket. The same structure does Tgs use to send the servticket in the fourth message.

9.2 Modelling Kerberos V

After a careful consideration of each of the novel working requirements outlined above, we decided not to include them in our formal model. It was a carefully biased decision. On the one hand, we found that the influence that

AUTHENTICATION

1. $A \rightarrow$ Kas : $A, \mathsf{Tgs}, T1$

2. Kas $\rightarrow A$: $\{\!\lvert authK, \mathsf{Tgs}, Ta \rvert\!\}_{Ka}, \underbrace{\{\!\lvert A, \mathsf{Tgs}, authK, Ta \rvert\!\}_{Ktgs}}_{authTicket}$

AUTHORISATION

3. $A \rightarrow$ Tgs : $\underbrace{\{\!\lvert A, \mathsf{Tgs}, authK, Ta \rvert\!\}_{Ktgs}}_{authTicket}, \underbrace{\{\!\lvert A, T2 \rvert\!\}_{authK}}_{authenticator}, B$

4. Tgs $\rightarrow A$: $\{\!\lvert servK, B, Ts \rvert\!\}_{authK}, \underbrace{\{\!\lvert A, B, servK, Ts \rvert\!\}_{Kb}}_{servTicket}$

SERVICE

5. $A \rightarrow B$: $\underbrace{\{\!\lvert A, B, servK, Ts \rvert\!\}_{Kb}}_{servTicket}, \underbrace{\{\!\lvert A, T3 \rvert\!\}_{servK}}_{authenticator}$

6. $B \rightarrow A$: $\{\!\lvert T3 + 1 \rvert\!\}_{servK}$

Fig. 9.1. Kerberos V protocol

these issues play on the main protocol goals, such as confidentiality and authentication (Chapter 4), was intuitive informal human reasoning. On the other side, a formal account of these issues appeared to be non-trivial. Formal models are notoriously only approximations to reality. Cryptography, for example, is notoriously reduced to a black-box primitive, thus neglecting differences such as the non-standard PCBC (Plain and Cipher Block Chaining) mode of DES used in Kerberos IV, and the more standard CBC mode used in the other version. Also, formal accounts are to be used particularly where human intuition falters. For example, we have seen the subtle outcomes of encrypting certain message components rather than leaving them in the clear with the Otway-Rees protocol (§8.5.1), and many similar cases can be found [58, 135]. In this vein, we shall see how difficult it would be to spot a certain lack of explicitness by informal reasoning on a smartcard protocol (§11.4).

By contrast, an informal account of the main design change, where the tickets are not doubly encrypted, is not trivial. On the one hand, we are currently aware that "extra encryption" is not necessarily the same as "extra security" [7, 14, 152]. On the other hand, not only did we prove the strong goals of Kerberos IV with double encryption, but used the same design strategy to somewhat strengthen the Otway-Rees protocol (§8.5.1). While it is clear that Otway-Rees can be strengthened exactly due to the use of extra encryption, it is not as clear to what extent double encryption determines the strong goals of Kerberos IV. Modelling the disposal of double encryption of Kerberos V is expected to help clarify this issue.

The protocol model `kerbV` can be found with file `KerberosV.thy` (Figure 3.1). As with the previous version, it is built on top of `Public.thy`, although it only uses symmetric cryptography. The inductive definition of `kerbV` resembles that of `kerbIV` seen above (§7.2) except for the formalisation of steps 2 and 4, which the postconditions of rules *KV2* and *KV4* respectively show in Figure 9.2.

KV2:
```
⟦ evs2 ∈ kerbV; Key authK ∉ used evs2; authK ∈ symKeys;
  Says A' Kas {|Agent A, Agent B, Number T1|} ∈ set evs2 ⟧
⟹ Says Kas A {|Crypt (shrK A) {|Key authK, Agent Tgs,
                                Number (CT evs2)|},
             Crypt (shrK Tgs) {|Agent A, Agent Tgs, Key authK,
                                Number (CT evs2)|}|}
      # evs2 ∈ kerbV
```

KV4:
```
⟦ evs4 ∈ kerbV; Key servK ∉ used evs4; B ≠ Tgs;
  authK ∈ symKeys; servK ∈ symKeys;
  Says A' Tgs {|Crypt (shrK Tgs) {|Agent A, Agent Tgs,
                                   Key authK, Number Ta|},
              Crypt authK {|Agent A, Number T2|}, Agent B|} ∈ set evs4;
  ¬ expiredAK Ta evs4; ¬ expiredA T2 evs4 ⟧
⟹ Says Tgs A {|Crypt authK {|Key servK, Agent B, Number (CT evs4)|},
             Crypt (shrK B) {|Agent A, Agent B, Key servK,
                              Number (CT evs4)|}|}
      # evs4 ∈ kerbV
```

Fig. 9.2. Inductive model of Kerberos V: fragment

Rules *KV3* and *KV5* are omitted from this presentation. They accommodate the absence of double encryption among their preconditions, corresponding to rules *KV2* and *KV4* respectively, so as to model reception of the right messages.

The current model rests on the formalisation of agents' knowledge based on trace inspection seen in the previous chapter, so there is no mention of the Gets event or of the knows function, but there is interest in the Issues properties. The usual formalisation of timestamping (§6.2) is retained. It can be seen that Tgs operates the temporal check advanced above (§7.3.5, condition 7.1) to make the goals of confidentiality, authentication and key distribution available to the peers in our threat model. It remains to be verified whether Kerberos V achieves the same results even with lighter use of encryption. This is the main subject of the next section.

9.3 Verifying Kerberos V

This section outlines the guarantees we have proved about Kerberos V. We skip everything that turned out to be in common with Kerberos IV, focusing only on the novel results. Here, *evs* is a generic trace of the formal protocol model kerbV. We follow our usual conventions (§1.3.2) to name the theorems.

9.3.1 Main Guarantees

The design variations of Kerberos V produce minor modifications to the authenticity and unicity arguments. Theorems analogous to Theorems 7.3.3 and 7.3.5 continue to assess authenticity of the authkey and of the servkey respectively (**KV_authK_authentic** and **KV_servK_authentic**, omitted here). However, they present existentially quantified tickets in the conclusions because the new message design does not let the assumptions bind the tickets. Fragments of the proof script pertaining to the unicity argument can be found in Appendix B.1.

We were pleased to observe that also the main guarantees proved about Kerberos IV — confidentiality, authentication and key distribution — substantially remain unvaried in the new version. Adherence to goal availability is maintained, ultimately confirming that the removal of double encryption has not weakened the protocol.

The confidentiality argument has merely to accommodate the variations of message forms. For example, this can be observed on the confidentiality guarantee for B on the servkey (Theorem 9.3.1), which is very similar to Theorem 7.3.14.

Theorem 9.3.1 (KV_Confidentiality_B). *If A and B are uncompromised, B is not* Tgs *and evs is such that*

Crypt(shrK B){Agent A, Agent B, Key $servK$, Number Ts}
    $\in$ parts(spies *evs*) *and*
Crypt $authK$ {Key $servK$, Agent B, Number Ts}
    $\in$ parts(spies *evs*) *and*
Crypt(shrK A){Key $authK$, Agent Tgs, Number Ta}
    $\in$ parts(spies *evs*) *and*
¬expiredAK Ta *evs* *and* ¬expiredSK Ts *evs*

then evs is such that

Key $servK$ $\notin$ analz(spies *evs*).

Observe that this version is for the original protocol, where Tgs does not operate our temporal check on expiry times. The corresponding version for the updated Kerberos V protocol (**KVu_Confidentiality_B**, omitted here) turns out to be identical to the theorem proved for the similarly updated

Kerberos IV protocol (**KIVu_Confidentiality_B**, §7.3.5). Both theorems in fact do not involve instances of the second or fourth message. This reasoning confirms that also Kerberos V in particular makes confidentiality of the servkey available to the responder in our threat model.

The same minor variations affect the authentication argument. Fragments of the relevant proof script can be found in Appendix B.3. For example, the combined guarantee of authentication of A with B and of key distribution of the servkey to A (Theorem 9.3.2) is, as we can expect, very similar to Theorem 8.4.4.

Theorem 9.3.2 (KV_A_authenticates&keydist_to_B). *If A and B are uncompromised, B is not* Tgs *and evs is such that*

> Crypt $servK$ (Number $T3$) $\in$ parts(spies evs)   *and*
>
> Crypt $authK$ {Key $servK$, Agent B, Number Ts}
>
> $\in$ parts(spies evs)   *and*
>
> Crypt(shrK A){Key $authK$, Agent Tgs, Number Ta}
>
> $\in$ parts(spies evs)   *and*
>
> Key $authK \notin$ analz(spies evs)   *and*   Key $servK \notin$ analz(spies evs)

then evs is such that

> B Issues A with (Crypt $servK$ (Number $T3$)) *on evs.*

The specular version (**KV_B_authenticates&keydist_to_A**, omitted here) turns out to be identical to the theorem proved for Kerberos IV (Theorem 8.4.3).

9.3.2 Novel Proof Methods

Despite our neat and simple conclusions, verifying Kerberos V was everything but a mere exercise. New proof methods had to be invented and developed, significantly beyond the mentioned simplification of terms involving the tickets.

For example, cryptic variables such as *authTicket* and *servTicket*, formalising the tickets, often appear throughout the proof process. They must be expanded, namely their form must be derived, for Isabelle's simplifier to be able to perform symbolic evaluation of terms involving the parts or analz operators. Otherwise, the simplifier would merely carry those variables along without performing any term reduction; the variables could potentially hide some message components, such as cryptographic keys, that would influence the simplification. The reader can endeavour in almost any proof attempt to obtain a demonstration of this observation. To show just one of innumerable examples, the subgoal concerning step 5 of the proof of **KV_servK_authentic** presents a term that demands simplification. It is

Crypt $authK$ {|Key $servK$, Agent B, Ts|}

∈ parts

    ({$servTicket$} ∪

        {Crypt $servK'$ {|Agent A, Number(CT $evs5$)|}} ∪

        (spies $evs5$))

Expanding the form of the tickets is easy with Kerberos IV when they are encrypted with confidential keys — respectively the initiator's shared key and an authkey. This was possible exactly thanks to the presence of double encryption. Conversely, it is impossible with Kerberos V for the opposite reason, because the message is concatenated. For example, let us consider the fourth message of Kerberos V. When the message

$$\{|\text{Crypt } authK \, \{|\text{Key } servK, \text{Agent } B, \text{Number } Ts|\}, servTicket|\} \qquad (9.1)$$

appears in the traffic, there is no way to derive the form of $servTicket$ because the message is concatenated; hence, the Spy might have replaced its second component with anything.

Fortunately, we could find an alternative and simple proof method to derive everything that is needed to know about the tickets for symbolic evaluation to get through. Here it is.

> Whenever a full instance of the second or of the fourth messages of Kerberos V appears in the traffic, then so does the mentioned ticket.

For example, this method applies to message 9.1. It is important to recall here that appearance in the traffic means that the message belongs to parts(spies evs), where evs is the current trace in the subgoal. The method can also be phrased in terms of analz(spies evs). Clearly, it also applies to authtickets whenever they appear in an instance of the second message. The full proof script about the protocol confirms that the lemmas implementing the method just described (**KV_Says_ticket_parts** and **KV_Says_ticket_analz**, omitted here) are vastly used.

Observe that this method can be partially used also with Kerberos IV, at least in terms of the parts operator. It would not be straightforward to use it in terms of the analz operator because this would require case studies on the keys encrypting the entire (second or fourth) messages.

The major proof obstacle derives from the attempt to prove an Issues property (Theorem 9.3.3).

Theorem 9.3.3 (KV_A_Issues_B). *If A and B are uncompromised, B is not* Tgs *and evs contains*

Says A B {|$servTicket$, Crypt $servK$ {|Agent A, Number $T3$|}|}

and is such that

Key $servK$ ∉ analz(spies evs)

then evs is such that

> A Issues B with (Crypt *servK* ⦃Agent A, Number *T3*⦄) on *evs*.

According to the general method (§8.1.2), this theorem is necessary to subsequently prove a combined guarantee of authentication and key distribution (**KV_B_authenticates&keydist_to_A**, omitted here). The proof scripts confirm that this can be done routinely with Kerberos V. However, the proof of Theorem 9.3.3 is more problematic than the corresponding one for Kerberos IV in part because of the mentioned existential conclusions of the authenticity theorems. Precisely, the subgoal corresponding to step 5 of the protocol is ultimately reduced as in Figure 9.3. It is difficult to see at first glance how to solve this subgoal.

```
⟦ A ∉ bad; B ∉ bad; A ≠ Kas; A ≠ Tgs; B ≠ Tgs; evs5 ∈ kerbV;
  authK ∈ symKeys; servK ∈ symKeys;
  Says Tgs' A ⦃Crypt authK ⦃Key servK, Agent B, Number Ts⦄, servTicket⦄
    ∈ set evs5;
  valid Ts wrt T2; Key servK ∉ analz (spies evs5);
  servTicket ∈ parts (spies evs5);
  Says A B ⦃servTicket, Crypt servK ⦃Agent A, Number (CT evs5)⦄⦄
    ∉ set evs5;
  Says Kas A ⦃Crypt (shrK A) ⦃Key authK, Agent Tgs, Ta⦄, authTicket⦄
    ∈ set evs5;
  Key authK ∉ analz (spies evs5);
  Says Kas A ⦃Crypt (shrK A) ⦃Key authK, Agent Tgs, Ta'⦄, authTicket'⦄
    ∈ set evs5;
  authTicket ∈ parts (spies evs5);
  Says Tgs A ⦃Crypt authK ⦃Key servK, Agent B, Number Ts⦄,
            Crypt (shrK B) ⦃Agent A, Agent B, Key servK, Number Ts⦄⦄
    ∈ set evs5;
  Crypt servK ⦃Agent A, Number (CT evs5)⦄ ∈ parts (spies evs5) ⟧
⟹ False
```

Fig. 9.3. Proving an Issues property for Kerberos V

When we proved an analogous guarantee (Theorem 7.3.17) for Kerberos IV, we appealed to a lemma **KIV_Says_K5** concluding that A sends the exact instance of the fifth message that would contradict the Says A B ... ∉ set *evs5* assumption and therefore solve the subgoal. However, an analogous lemma that can be proved for Kerberos V carries a crucial existential quantifier upon the copy of the servTicket in the conclusion (Lemma 9.3.1).

Lemma 9.3.1 (KV_Says_K5). *If A and B are uncompromised and evs contains*

> Says Tgs A ⦃Crypt *authK* ⦃Key *servK*, Agent B, Number *Ts*⦄, *servTicket*⦄

and is such that

Crypt $servK$ {|Agent A, Number $T3$|} $\in$ parts(spies evs)   *and*

Key $authK$ $\notin$ analz(spies evs)

then, for some sT, evs contains

Says $A\,B$ {|sT, Crypt $servK$ {|Agent A, Number $T3$|}|}.

This lemma cannot be strengthened. Because the fourth message is no longer doubly encrypted, there exists no guarantee that the servTicket that Tgs issues for A is not altered during transmission to A. Because that ticket is unintelligible for A, she merely has to forward it to B. In consequence, A will forward anything (of the right length) that the Spy might have inserted instead of the original ticket. The lemma inevitably reflects this by existentially quantifying variable sT in its conclusion. That variable cannot be stated to necessarily match *servTicket*.

Applying Lemma 9.3.1 to the subgoal described above indicates that the current trace *evs5* also contains the event

$$\text{Says } A\,B \ \{|servTicket', \text{Crypt } servK \ \{|\text{Agent } A, \text{Number}(\text{CT } evs5)|\}|\} \quad (9.2)$$

but it is impossible to prove that *servTicket* equals *servTicket'*, which would let us conclude.

Significant experiments — most of which failed! — were conducted at this stage to find an alternative proof method. Finally, we realised that a novel, meticulous reasoning on timestamps could help. The new method is based on the fundamental observation that uncompromised agents always insert the right timestamps in the messages they form. Each timestamp is taken by an uncompromised agent as the current time of the trace. In consequence, an uncompromised agent never issues a timestamp before the right time for it.

To demonstrate these considerations, Figure 9.4 shows a simple protocol trace that only records an uncompromised agent A running the protocol up to step 2. Recall that traces are lists of protocol events and that are built in reverse chronological order (§3.8). The example trace begins with the event formalising step 1 of the protocol and continues with the event formalising step 2. This version of the protocol model does not include message reception; otherwise, the appropriate reception event would have appeared between the two given events.

```
Says Kas A {Crypt (shrK A) {Key authK, Agent Tgs, Number 1},
           Crypt (shrK Tgs) {Agent A, Agent Tgs, Key authK, Number 1}}

Says A Kas {Agent A, Agent Tgs, Number 0}
```

Fig. 9.4. Example trace belonging to the inductive model of Kerberos V

Observe that the trace is initially empty; hence, A picks 0 as the current time to insert as the timestamp in the first event. Likewise, Kas picks 1 as the current time of the trace with the first event only, and inserts it as the timestamp in the second event. The assumption that A is uncompromised guarantees that, if she continues this trace with the event formalising step 3, she will insert 2 as the timestamp in that event. Clearly, the Spy behaves arbitrarily; hence, she can insert any timestamp, past or future, in any event.

Following these considerations, we can prove that an uncompromised agent always inserts the right timestamp in an authenticator, namely that he cannot have inserted t as a timestamp in a trace whose length is exactly t (Lemma 9.3.2).

Lemma 9.3.2 (KV_honest_never_says_current_timestamp_in_auth).
If A is uncompromised, evs and X are such that

$$\mathsf{Number}(\mathsf{CT}\ evs) \in \mathsf{parts}\{X\}$$

then evs does not contain

$$\mathsf{Says}\,A\,B\,\{\!|Y,X|\!\}.$$

It is interesting to observe that this lemma cannot be proved by induction as it stands. A variant can be proved, where the main assumption is replaced by

$$\mathsf{Number}\,T \in \mathsf{parts}\{X\}\quad\text{and}\quad(\mathsf{CT}\ evs) \leq T$$

Lemma 9.3.2 is a particular case of this variant, which can be proved by induction, then simplification and finally the `force` prover. Observe that the result applies to both steps 2 and 4, X being a concatenated message only in the latter case. We have also investigated whether the conclusion could be strengthened as $\mathsf{Says}\,A\,B\,X$. A failed proof attempt indicated that the Spy could replace a ticket, while it is in the network, with a past timestamp. An uncompromised receiver would merely forward it as the first component of the third or the fifth message, and thus falsify the putative conclusion. By contrast, we verified that the stronger lemma would hold for Kerberos IV thanks exactly to the use of double encryption, which protects the tickets, if B's refresh of the timestamp sent in the last step is modelled.

Going back to the proof of Theorem 9.3.3, the application of Lemma 9.3.2 produces with event 9.2 the contradiction that can solve the subgoal in Figure 9.3.

9.3.3 Novel Guarantees

The novel proof methods presented in the previous section rely on timestamps. We have seen that an uncompromised agent always picks the right timestamp to insert into the message he is creating. That timestamp is the current time of the trace, not an earlier (or a later) time; hence, it is fresh.

On the basis of these novel considerations, we argue that a unicity guarantee must hold for each event that sees an uncompromised agent send a message containing a timestamp just chosen. We have developed new guarantees to support this argument. Fragments of the relevant proof script can be found in Appendix B.2.

For example, a unicity guarantee of the original form holds for the first authenticator (Theorem 9.3.4), which is sent in the first step of the protocol.

Theorem 9.3.4 (KV_unique_timestamp_authenticator1). *If A is uncompromised and evs contains*

$$\text{Says } A \text{ Kas } \{\!|\text{Agent } A, \text{Agent Tgs}, \text{Number } T1|\!\}$$

and

$$\text{Says } A \text{ Kas}' \{\!|\text{Agent } A, \text{Agent Tgs}', \text{Number } T1|\!\}$$

then

$$\text{Kas} = \text{Kas}' \quad and \quad \text{Tgs} = \text{Tgs}'.$$

Technically speaking it would have been sufficient to merely assume A not to be the Spy. The proof is a simple induction followed by Isabelle's combined use of simplification and classical reasoning, the auto method. Only one subgoal, corresponding to the first step of the protocol, survives. It can be solved by an invocation to Lemma 9.3.2 through the blast method.

As can be expected, analogous guarantees hold for the second authenticator (**KV_unique_timestamp_authenticator2**, omitted here) and for the third (**KV_unique_timestamp_authenticator3**, omitted here), which are respectively addressed to Tgs and to the required service B.

The same proof methods apply to the messages issued by Kas or Tgs, in which the two Servers, which are assumed to operate reliably, insert fresh timestamps. Precisely, Kas never uses the same timestamp within two distinct message contexts (Theorem 9.3.5).

Theorem 9.3.5 (KV_unique_timestamp_authTicket). *If evs contains*

$$\text{Says Kas } A \{\!|X, \text{Crypt}(\text{shrK Tgs})\{\!|\text{Agent } A, \text{Agent Tgs}, \text{Key } authK,$$
$$\text{Number } Ta|\!\}|\!\}$$

and

$$\text{Says Kas } A' \{\!|X', \text{Crypt}(\text{shrK Tgs}')\{\!|\text{Agent } A', \text{Agent Tgs}', \text{Key } authK',$$
$$\text{Number } Ta|\!\}|\!\}$$

then

$$A = A' \quad and \quad X = X \quad and \quad \text{Tgs} = \text{Tgs}' \quad and \quad authK = authK'.$$

An analogous guarantee (**KV_unique_timestamp_servTicket**, omitted here) holds for the message that Tgs issues. Both theorems invoke Lemma 9.3.2 although the second parts of the messages that Kas or Tgs send are not authenticators. The assumptions of the lemma in fact become available during their proofs. For example, let us consider the proof of Theorem 9.3.5. After induction, a single invocation of the auto method enriched with Lemma 9.3.2 as a simplification rule terminates the proof. Alternatively, to inspect the proof more closely, we may take the following analytic steps after induction.

- Simplify all subgoals through method simp_all. It terminates all subgoals except the one concerning step 2, which is non-trivial.
- Simplify the subgoal appealing to Lemma 9.3.2 through method simp add followed by the name of the lemma. It simplifies away a few contradictory cases from the conclusion.
- Clarify the subgoal through method clarify. It performs all obvious reasoning steps without splitting the goal into multiple parts. The resulting subgoal can be seen in Figure 9.5.
- Another application of Lemma 9.3.2 contradicts the Says Kas A' ... $\in$ set evs2 assumption because the second component of the message is a ciphertext containing exactly the timestamp CT evs2. The proof terminates.

```
⟦ evs2 ∈ kerbV; Key AK ∉ used evs2; AK ∈ symKeys;
  Says A' Kas ⦃Agent A, Agent Tgs, Number T1⦄ ∈ set evs2;
  Says Kas A'
    ⦃X', Crypt (srhK Tgs')
           ⦃Agent A', Agent Tgs', Key AK', Number (CT evs2)⦄⦄
  ∈ set evs2;
  Says Kas A
    ⦃Crypt (shrK A) ⦃Key AK, Agent Tgs, Number (CT evs2)⦄,
       Crypt (shrK Tgs) ⦃Agent A, Agent Tgs, Key AK, Number (CT evs2)⦄⦄
  ∉ set evs2⟧
⟹ A = A' ∧
    Crypt (shrK A) ⦃Key AK, Agent Tgs, Number (CT evs2)⦄ = X' ∧
    Tgs = Tgs' ∧ AK = AK'
```

Fig. 9.5. Proving a novel unicity property for Kerberos V

Moreover, it can be observed that the two Servers in fact insert two occurrences (rather than just one) of a fresh timestamp in each of their messages. Therefore, appropriate unicity guarantees can be proved using the first occurrence as a pivot. We only present the guarantee for Tgs (Theorem 9.3.6), while an analogous one (**KV_unique_timestamp_msg2**, omitted here) holds for Kas.

Theorem 9.3.6 (KV_unique_timestamp_msg4). *If evs contains*

Says Tgs A {Crypt(shrK A){Key $servK$, Agent B, Number Ts},
  $serv\,Ticket$}

and

Says Tgs A' {Crypt(shrK A'){Key $servK'$, Agent B', Number Ts},
  $serv\,Ticket'$}

then

$A = A'$ *and* $servK = servK'$ *and* $B = B'$ *and*
$serv\,Ticket = serv\,Ticket'$.

Also for these theorems the proof method is substantially unvaried. However, they require inspection of the first component of the message that is sent, and hence cannot appeal to Lemma 9.3.2, which inspects the second component. However, it is easy to prove a similar lemma for the generic message sent by either of the two Servers (Lemma 9.3.3).

Lemma 9.3.3 (KV_Servers_never_says_current_timestamp). *If evs and X are such that*

Number(CT evs) $\in$ parts{X}

then evs does not contain

Says Kas $A\,X$ *or*   Says Tgs $A\,X$.

The theorems presented here are only of technical relevance. They also express a form of reliability of the protocol model; in fact, similar forms are expected to hold for Kerberos IV and for any protocol that requires an agent to insert the current time as a timestamp. Also, they can be reformulated in terms of the Unique predicate to convey more stringent guarantees.

Another technical observation is that, while we have formally proved that an uncompromised agent does not send the same timestamp more than once in a trace, there is nothing to prevent him from sending the same timestamp on two different traces of the same length. This is to be ascribed to our formalisation of the current time of a trace as the length of the trace. The same remark applies to fresh nonces or session keys, which are guaranteed to be issued only once in a trace.

It was thanks to the novel design of Kerberos V, which required the investigation of novel proof methods, that we could conceive the new guarantees. Their formal statements were not obvious to design, but their proofs turned out to be reasonably easy.

10. Modelling Smartcards

The Inductive Method is tailored to the analysis of protocols that make use of smartcards. The threat model is extended by giving the Spy the ability to clone other agents' cards and exploit their computational resources.

The protection of long-term secrets was the main reason for the introduction of smartcards. Although several researchers believe that no smartcard can be completely tamper resistant, modern cards offer a high level of physical security. A cheap *integrated-circuit memory card* may store a few kilobytes and provide a strong shell for important information. Additionally, an *integrated-circuit microprocessor card* embeds an up to 32-bit (or 64-bit RISC) microprocessor. It can perform relatively simple operations such as DES or RSA encryption/decryption with relatively long keys. Consequently, existing security protocols have been extended with smartcards [95] and new ones have been designed relying on these inexpensive devices [148]. We refer to the protocols that are based on smartcards as *smartcard protocols*, as opposed to the *traditional protocols*, which are not.

The modern operating systems of microprocessor cards allow the execution of user-chosen Java programs, cutting the costs of applications such as pay TV, mobile or public phones and credit cards. It is difficult to estimate the number of cards that exist worldwide. Each mobile phone contains one. VISA estimated as 42 million the total number of Visa Corporation branded smartcards in 2001 [162]. It is certain that not only the actual cards but also the smartcard readers have become inexpensive pieces of hardware for home computers running smartcard middleware.

Smartcards should strengthen the goals of the protocols that use them, and there exists an increasing demand for formal guarantees that this target is indeed reached. Two are the most significant attempts to establish such guarantees. Abadi et al. [3] pioneer the use of smartcards to establish mutual authentication between agents and workstations. Their treatment develops around the different functionalities of the adopted smartcards, while facing the limited technology of the time, and is based on a belief logic that is a simple extension of the BAN logic (§2.1.2). The calculus of the logic is used to prove the mutual authentication and delegation goals of three protocols that impose different requirements on the cards. Confidentiality issues are not considered, as is common with belief logics. Shoup and Rubin [148] design a protocol based on smartcards and analyse it using a complexity-theoretic method based on provable security (§2.1.4).

Motivated by the insufficient research done in the field, we extend the Inductive Method towards the verification of smartcard protocols [26]. The model cards, which are associated with a new type of the formal language, can interact with their respective owners by receiving and sending messages. Each card stores a basic set of long-term secrets, which may depend on the specific protocol. For example, while a card for key distribution protocols has only to store specific keys, a card for e-commerce protocols may also have to store a number representing the owner's balance. The cards are not forced to perform any computations and may skip some or repeat others. The Spy has stolen an unspecified set of cards but must discover their pins, if they are PIN operated, to be able to use them. Furthermore, she has cloned another set of cards, discovering their internal secrets. So, since the Spy can act illegally, there is a set of cards that she can use even if they do not belong to her, while all other agents can only use their own card.

Several smartcard protocols make the *assumption of secure means*, signifying that the Spy cannot interpose between agents and their cards. So, messages can be exchanged in the clear and each agent's knowledge of long-term secrets reduces to nothing. We account for both this and the opposite alternative by simple variations to the definition of Spy's knowledge. Often in the following text, *secure means* will abbreviate that the assumption of secure means holds; *insecure means* will abbreviate that the assumption does not hold.

This chapter begins with a formalisation of smartcards within the Inductive Method (§10.1). Then, it presents the extensions necessary to the datatype of events (§10.2) and to the definition of agents' knowledge (§10.3). The Spy's illegal behaviour now exploits certain smartcards that she does not legally own (§10.4), while the protocol model may require some extensions to account for this (§10.5). This extended approach will be demonstrated in the next chapter.

10.1 Smartcards

The treatment presented in this chapter is derived from files EventSC.thy and Smartcard.thy (Figure 3.1). They respectively contain the events necessary to model a smartcard protocol and a specification of smartcards [33, 34].

To represent the operational aspects of smartcards, a new free type card is introduced with several associated functions. A bijective correspondence

Card : agent $\longrightarrow$ card

is stated between agents and smartcards, formally ruling out the chance that an agent owns more than one card. This is a choice of simplicity. Should that chance become reality, the single model card could be imagined to incorporate all computational resources. Alternatively, the same function can map an agent into a set of cards.

In the real world, the card CPUs only provide certain, limited resources: a card will produce a specific output only if fed the correct input. For example, if a card can compute a session key K from an input X, the card must necessarily be fed X in order to obtain K. The formal protocol model can easily account for this. It will allow for the outputs encompassed by the protocol only under the condition that the cards are fed the corresponding, specific inputs. It will not construct other outputs, even from cloned cards (§10.1.1). In consequence, there exists no card whose use can give the Spy unlimited power.

10.1.1 Card Vulnerabilities

The model cards suffer from a number of realistic vulnerabilities due to theft, cloning and internal failures.

Theft. The small dimensions of the smartcards confer their portability but also raise the risk of loss or theft. In the worst case, all smartcards that have been lost by their owners or stolen from them will end up in the Spy's hands. These cards, which can no longer be used by their owners, are modelled by the set stolen, such that stolen $\subseteq$ card.

Cloning. The Spy is not necessarily able to use a stolen card actively, unless she knows its PIN. Nevertheless, she could be able to use modern techniques (such as *microprobing* [15]), break the physical security of the card, access its EEPROM (namely Electrically Erasable Programmable Read-Only Memory) where the long-term secrets are stored and, in the worst case, reverse engineer the whole card chip. At this stage, the Spy would be able to build a clone of the card for her own use. If this process succeeds, the card belongs to the set cloned in the model, and cloned $\subseteq$ card.

Cloning without apparent theft. All cloning techniques that are currently known are *invasive*, in the sense that they spoil the original card. The chip of the card must be disembedded from its frame by suitable chemicals, and its layout often modified using laser cutter microscopes. These alterations are irreversible. However, the Spy might steal a card, build two clones of it and return one to the card owner, who would not suspect anything. Alternatively, the Spy might even be able to tailor *non-invasive* techniques (such as *fault generation* [103, 113] by exploiting the power and clock supply lines) to cloning in the near future, and return the original card to its owner after building a clone for herself. Modelling these opportunities simply requires stating no relation between the sets stolen and cloned, so that a card could be cloned and not be stolen. Such a card could be used both by its legal owner and by the Spy, granting them identical computational resources.

Data bus failure. All card data buses are corrupted so that the travelling messages can be either forgotten, for example due to electronic decay, permuted or fed to the CPU repeatedly, for example due to simple layout modifications. Allowing message alteration or leakage at this level would give the Spy excessive, unrealistic power: many protocols explicitly rely on secure means between agents and smartcards. Therefore, a smartcard can omit some computations or repeat others. In the worst case, the Spy has caused all cards to deteriorate in this fashion before they are delivered to their respective owners, leaving no visible trace of tampering. This is easy to model inductively: events only occur by firing of inductive rules in the formal protocol model, but rules are not forced to fire even when their preconditions are met. Also, rules may fire in any order or fire more than once, provided that their preconditions are met: each of these possibilities is recorded by a specific trace.

Global internal failure. Smartcards may suffer unexpected failures and abruptly stop working at any point. A trace of the formal protocol model easily reflects each scenario of global internal failure. Some events mentioning the card can be identified in the trace. These events terminate at some stage, namely there exists an event mentioning the card such that the continuation of the trace does not feature the card any more. It reflects the permanent exclusion of the card from the protocol during the very history that the trace models.

10.1.2 Card Usability

Agents other than the Spy only conduct legal operations, while the Spy can act both legally and illegally. A card that has not been stolen can be used by its owner, namely it can be used legally. The Spy cannot handle an unstolen card unless it is her own.

Definition 10.1.1. $\mathsf{legalUse}(\mathsf{Card}\,A) \triangleq \mathsf{Card}\,A \notin \mathsf{stolen}.$

When the assumption of secure means does not hold, the Spy can listen in between any agent and his smartcard; so, she has electronic access to those cards of which she knows the pins (if a smartcard is PIN operated, then it accepts no communication unless it is activated by means of its PIN). Pins are sent between agents and cards (never between agents), so the Spy might learn some of them on certain traces. By contrast, should the cards not be PIN operated, they would all be illegally usable.

Definition 10.1.2. Let the assumption of secure means not hold;

$$\mathsf{illegalUse}(\mathsf{Card}\,A)\ \text{on}\ evs \triangleq \begin{cases} the\ Spy\ knows\ A\text{'}s\ PIN\ on\ evs \\ \quad \textit{if cards are PIN operated} \\ true \\ \quad \textit{if cards are not PIN operated} \end{cases}$$

The informal predicate *the Spy knows A's PIN on evs* will be refined below by the formal definition of agents' knowledge (§10.3).

When the assumption of secure means holds, the Spy needs to gain physical access to the cards in addition to the knowledge of their pins. Since she cannot monitor the events involving the smartcards, she has no chance of discovering any pins via any events. She can only know them initially (§10.3), so the definition of illegal usability does not depend in the trace. If the cards are not PIN operated, we only need to characterise the physical access to the card.

Definition 10.1.3. Let the assumption of secure means hold;

$$
\mathsf{illegalUse}(\mathsf{Card}\,A) \triangleq
\begin{cases}
\mathsf{Card}\,A \in \mathsf{cloned} \;\;\text{or}\;\; (\mathsf{Card}\,A \in \mathsf{stolen} \;\;\text{and} \\
\qquad\qquad\qquad\qquad\quad \text{the Spy knows A's PIN}) \\
\qquad\qquad\qquad \text{if cards are PIN operated} \\[4pt]
\mathsf{Card}\,A \in \mathsf{cloned} \;\;\text{or}\;\; \mathsf{Card}\,A \in \mathsf{stolen} \\
\qquad\qquad\qquad \text{if cards are not PIN operated}
\end{cases}
$$

Also the informal predicate *the Spy knows A's PIN* will be refined below.

The Spy must be able to use her own card legally because she must be given the opportunity to act legally. However, she does not need to use her card illegally because she cannot acquire additional knowledge from doing so. It can be assumed that

$$\mathsf{Card}\,\mathsf{Spy} \notin \mathsf{stolen} \cup \mathsf{cloned}$$

The same is assumed of the card that belongs to the Server.

It must be stressed that, since certain cards may be cloned and at the same time not be stolen, there may exist cards that are both legally and illegally usable. Every agent is able to verify whether or not his own card is stolen by checking whether or not he holds it. All other assumptions about the agent's card or the agent's peer's card now belong to the minimal trust (Chapter 5).

10.1.3 Card Secrets

A smartcard typically contains two long-term symmetric keys: the PIN to activate its functionalities and the card key. As can be seen in file `Smartcard.thy` (Figure 3.1), we declare two corresponding functions

$$\mathsf{PIN} : \mathsf{agent} \longrightarrow \mathsf{key}$$

$$\mathsf{crdK} : \mathsf{card} \longrightarrow \mathsf{key}$$

The first of the two functions can be equivalently defined on cards rather than on agents. The card key serves to limit the data that must be stored in the card RAM. Suppose that when the card is required to issue a fresh nonce it also outputs the nonce encrypted under its key. This cipher may be used later to assess the nonce authenticity to the card, even if the card did not store the nonce, assuming that the card key is secure.

In the case of key distribution protocols, each card also stores its owner's long-term key, which is not known to the agent, in contrast with traditional protocols. We keep the original definition (§3.4)

shrK : agent $\longrightarrow$ key

Observe that, since the model is operational, the notion that the smartcards *store* some secrets does not need to be formalised explicitly. We only need to define how these secrets are used, namely in which circumstances and to whom they will become known. This increases the flexibility of the specification method. If the smartcards store additional secrets in certain applications, once such secrets are formalised by suitable functions, only the definition of agents' knowledge must be updated.

We assume that collision of keys is impossible, so all functions declared above are injective and their ranges are disjoint.

10.2 Events

The treatment presented here comes with the file `EventSC.thy` (Figure 3.1), which formalises the events for smartcard protocols. The theory is built on the standard `Message.thy` theory file; hence, it branches the theory hierarchy in parallel with file `Event.thy`, which contains the events for standard protocols (§8.2).

We introduce seven events for smartcard protocols that do not make the assumption of secure means. Conversely, two of them should be pruned over secure means, although we will retain them all in practice and merely require the protocol model to only use the right subset. The Isabelle datatype of events is upgraded as

datatype event $\triangleq$ Says agent agent msg
               Notes agent msg
               Gets agent msg
               Inputs agent card msg
               Gets_c card msg
               Outputs card agent msg
               Gets_a agent msg

The known *network events* (sending, noting and receiving a message, §8.2) have been extended with the new *card events*. Agents may send inputs to the cards (Inputs) and the cards may receive them (Gets_c); similarly, the cards may send outputs to the agents (Outputs) and the agents may receive them

(Gets_a). An agent can distinguish the messages received from the network from those received from his smartcard reader because they arrive on separate ports; so we provide two different events. However, in both cases the messages could have been forged by the Spy.

Extending the reception invariant (§8.2), the protocol model allows the cards to receive by a Gets_c event only the messages that have been sent by an Inputs event, and allows the agents to receive by a Gets_a event only the messages that have been sent by an Outputs event (§10.5).

When the assumption of secure means holds, the events Gets_c and Gets_a can be omitted: a card certainly receives its owner's inputs, and an agent certainly receives his card outputs. Consequently, a smartcard C can verify whether an event Inputs $A\,C\,X$ occurred and an agent A can verify whether an event Outputs $C\,A\,X$ occurred, while this is impossible on insecure means.

The formal definition of the function used (§§3.10 and 8.2) must be enriched with four rules to account for the new events.

4. All components of a message that an agent sends as input to a smartcard in a trace are used on that trace.

   used$((\mathsf{Inputs}\,A\,C\,X)\,\#\,evs) \triangleq \mathsf{parts}\{X\} \cup \mathsf{used}\ evs$

5. All messages that a smartcard receives as inputs from an agent in a trace do not directly extend those that are used on that trace.

   used$((\mathsf{Gets\_c}\,C\,X)\,\#\,evs) \triangleq \mathsf{used}\ evs$

6. All components of a message that a smartcard sends as output to an agent in a trace are used on that trace.

   used$((\mathsf{Outputs}\,C\,A\,X)\,\#\,evs) \triangleq \mathsf{parts}\{X\} \cup \mathsf{used}\ evs$

7. All messages that an agent receives as outputs from a smartcard in a trace do not directly extend those that are used on that trace.

   used$((\mathsf{Gets\_a}\,A\,X)\,\#\,evs) \triangleq \mathsf{used}\ evs$

Cases 5 and 7 do not extend the set of used components because the corresponding events pertain to messages already considered by means of the reception invariant. If the assumption of secure means holds, both cases are omitted.

10.3 Agents' Knowledge

The function initState formalising the agents' initial knowledge (§3.9) must be redefined to account for the secrets stored in the smartcards. We quote here a fairly general definition for smartcard protocols relying on the original datatype of agents (§3.3) and with PIN operated cards. Asymmetric long-term keys are omitted for brevity, while the ' symbol as usual indicates the image operator.

1. The Server's initial knowledge consists of all long-term secrets.

   initState Server $\triangleq$ (Key` range PIN) $\cup$

   (Key` range crdK) $\cup$ (Key` range shrK)

2. Each friendly agent's initial knowledge consists of his own PIN.

   initState (Friend i) $\triangleq$ {Key (PIN (Friend i))}

3. The Spy's initial knowledge consists of the compromised agents' secrets and of the secrets stored in the cloned cards (even if some cards store the secrets in a blinded form, the Spy discovers them in the worst case).

   initState Spy $\triangleq$ (Key` PIN` bad) $\cup$ (Key` PIN`{A. Card $A \in$ cloned}) $\cup$

   (Key` crdK` cloned) $\cup$

   (Key` shrK`{A. Card $A \in$ cloned})

Observe that this definition is not influenced by the assumption of secure means because it formalises the situation before any protocol sessions have taken place. The opposite applies to the knowledge that agents can extract from traces, which we define by the function knows (§8.2.1). For simplicity, we present here only the most interesting case, which occurs when the assumption of secure means does not hold. However, its definition, as is released in file EventSC.thy, can be found in Appendix C.2 along with a few technical lemmas. It is parametric over the flag secureM, whose truth value a protocol model merely has to state.

4. An agent knows what he inputs to any card in a trace; in particular, the Spy also knows all messages ever input on it.

   knows A ((Inputs A' C X) # evs) $\triangleq$

   $\begin{cases} \{X\} \cup \text{knows } A \; evs & \text{if } A = A' \text{ or } A = \text{Spy} \\ \text{knows } A \; evs & \text{otherwise} \end{cases}$

5. No agent, including the Spy, can extend his knowledge with any of the messages received by any smartcard in a trace. The Spy and the message originators already know them by case 4 thanks to the reception invariant.

   knows A ((Gets_c A' X) # evs) $\triangleq$ knows A evs

6. An agent knows no card outputs in a trace, as the means is insecure; the Spy knows all of them, as she controls the means.

   knows A ((Outputs C A' X) # evs) $\triangleq$

   $\begin{cases} \{X\} \cup \text{knows } A \; evs & \text{if } A = \text{Spy} \\ \text{knows } A \; evs & \text{otherwise} \end{cases}$

7. An agent other than the Spy knows what he receives from his card in a trace. The Spy knows all messages received by any smartcard by case 6, due to the reception invariant.

knows $A\,((\mathsf{Gets\_a}\,A'\,X)\,\#\,evs)\ \triangleq$

$$\begin{cases} \{X\} \cup \mathsf{knows}\,A\ evs & \text{if } A = A' \text{ and } A \text{ is not the } \mathsf{Spy} \\ \mathsf{knows}\,A\ evs & \text{otherwise} \end{cases}$$

At this stage, definition 10.1.2 can be refined. Recall that the function analz extracts all message components from a set of messages using keys that are recursively available (§3.10).

Definition 10.1.2′. Let the assumption of secure means not hold;

$$\mathsf{illegalUse}(\mathsf{Card}\,A)\text{ on } evs\ \triangleq\ \begin{cases} \mathsf{Key}\,(\mathsf{PIN}\,A) \in \mathsf{analz}(\mathsf{knows}\,\mathsf{Spy}\ evs) \\ \qquad \text{if cards are PIN operated} \\ \mathsf{true} \\ \qquad \text{if cards are not PIN operated} \end{cases}$$

In particular, a cloned card or a card whose owner is compromised is illegally usable on any trace (by definition of initState, base case of knows and definition of analz). As expected, the illegal usability of a card over insecure means does not necessarily imply the Spy's physical access to the card.

If the assumption of secure means holds, the definition of knows simplifies. The base case and those corresponding to the network events remain unchanged. Cases (5) and (7) must be pruned, for the corresponding events are no longer defined. If an agent sends an input to his card, or the card sends him back an output, both messages are certainly received because the Spy cannot listen in. Hence, cases (4) and (6) must be amended accordingly.

4′. An agent, including the Spy, knows what he inputs to any card in a trace.

knows $A\,((\mathsf{Inputs}\,A'\,C\,X)\,\#\,evs)\ \triangleq$

$$\begin{cases} \{X\} \cup \mathsf{knows}\,A\ evs & \text{if } A = A' \\ \mathsf{knows}\,A\ evs & \text{otherwise} \end{cases}$$

6′. An agent, including the Spy, knows what he is output from any card in a trace.

knows $A\,((\mathsf{Outputs}\,C\,A'\,X)\,\#\,evs)\ \triangleq$

$$\begin{cases} \{X\} \cup \mathsf{knows}\,A\ evs & \text{if } A = A' \\ \mathsf{knows}\,A\ evs & \text{otherwise} \end{cases}$$

As was desired, these cases forbid the Spy from learning anything from the card events. Therefore, she knows a PIN if and only if she knows it initially. By definition of initState and base case of knows, definition 10.1.3 can be refined as follows.

Definition 10.1.3′. Let the assumption of secure means hold;

$$\text{illegalUse}(\text{Card } A) \triangleq \begin{cases} \text{Card } A \in \text{cloned or } (\text{Card } A \in \text{stolen and } A \in \text{bad}) \\ \qquad\qquad\qquad\qquad\qquad\text{if cards are PIN operated} \\ \text{Card } A \in \text{cloned or } \text{Card } A \in \text{stolen} \\ \qquad\qquad\qquad\qquad\text{if cards are not PIN operated} \end{cases}$$

This definition insists on the Spy's physical access to the illegally usable cards over secure means. Only when the cards are not PIN operated over secure means does it hold that if a card is not illegally usable, then it is legally usable. This does not hold in general, nor does the converse.

The function knows can be extended to smartcards, but a detailed reasoning about card knowledge may seem exaggerated at present due to their limited RAM. However, the next chapter at times advances some considerations in this direction.

10.4 Threat Model

We have seen above (Chapters 3, 6 and 7) that the threat model for traditional protocols is typically specified by the single inductive rule *Fake* extending the protocol model.

In modelling smartcard protocols, the Spy must be allowed to exploit the illegally usable smartcards. If the assumption of secure means does not hold, not only can the Spy send fake messages as inputs to the illegally usable cards, but she can also send fake outputs to any agents, pretending that her own card could produce them. This is done in addition to sending the fake messages on the network because receiving the same message from the card reader or from the network may induce different reactions in an agent. The *Fake* rule must be amended as outlined in Figure 10.1. Observe the condition of illegal usability over insecure means stated on A's card.

```
Fake :
⟦ evsF ∈ smart_p_insecure_m; illegalUse(Card A) on evs;
   X ∈ synth (analz (knows Spy evsF)) ⟧
⟹ Says Spy B X # Inputs Spy (Card A) X # Outputs (Card Spy) C X
      # evsF ∈ smart_p_insecure_m
```

Fig. 10.1. Rule template for Spy's illegal behaviour in the case of insecure means

If the assumption of secure means holds, then the Spy cannot send fake card outputs to the agents. Figure 10.2 presents the corresponding, new *Fake* rule. Observe the condition of illegal usability over secure means stated on A's card.

In this scenario, by definition of knows, the Spy gains no knowledge from the card events that do not concern her. Therefore, we must assure that an

```
Fake:
⟦ evsF ∈ smart_p_secure_m; illegalUse(Card A);
   X ∈ synth (analz (knows Spy evsF)) ⟧
⟹ Says Spy B X # Inputs Spy (Card A) X
        # evsF ∈ smart_p_secure_m
```

Fig. 10.2. Rule template for Spy's illegal behaviour in the case of secure means

illegally usable card outputs towards the Spy rather than towards its owner. This is realistic because, over insecure means, the illegally usable cards lie in the Spy's hands. Suppose that A's card outputs X' when it is fed X. The formal protocol model will contain rule *Name* (Figure 10.3), which requires the card to be legally usable. Hence, rule *Name_Fake* must be added to allow A's card to output X' towards the Spy in case the card is illegally usable and was input X by the Spy. The Spy learns X' from the firing of the latter rule, not of the former. Any extra assumptions in *Name* must be kept in *Name_Fake*. Should A's card be both legally and illegally usable, both rules would be enabled to fire. Rule *Name_Fake* is unnecessary over insecure means, where the Spy monitors all card events.

```
Name:
⟦ evsN ∈ smart_p_secure_m; legalUse(Card A);
   Inputs A (Card A) X ∈ set evsN ⟧
⟹ Outputs (Card A) A X' # evsN ∈ smart_p_secure_m

Name_Fake:
⟦ evsNF ∈ smart_p_secure_m; illegalUse(Card A);
   Inputs Spy (Card A) X ∈ set evsNF ⟧
⟹ Outputs (Card A) Spy X' # evsNF ∈ smart_p_secure_m
```

Fig. 10.3. Rule templates for each card output in the case of secure means

10.5 Protocol Model

The formal model for a smartcard protocol requires additional features if the assumption of secure means does not hold.

The smartcards must be allowed to receive the inputs that they were sent from agents and, likewise, agents must be allowed to receive the outputs sent from cards. For the respective purposes, we introduce in Figure 10.4 rules *Reception_c* and *Reception_a*, which are inspired by the *Reception* rule for messages sent over the network (§8.2.2). Since the rules are not forced to fire,

```
Reception_c:
⟦ evsRc ∈ smart_p_insecure_m; Inputs A (Card B) X ∈ set evsRc ⟧
⟹ Gets_c (Card B) X # evsRc ∈ smart_p_insecure_m

Reception_a:
⟦ evsRa ∈ smart_p_insecure_m; Outputs (Card A) B X ∈ set evsRa ⟧
⟹ Gets_a B X # evsRa ∈ smart_p_insecure_m
```

Fig. 10.4. Rule templates for message reception in the case of insecure means

no kind of reception (either from the network or from the agent-smartcard means) is guaranteed, as is the case in a world where the Spy controls all means.

If the assumption of secure means holds, then reception over the agent-smartcard means is guaranteed, so the rules in Figure 10.4 are not needed.

11. Verifying a Smartcard Protocol

The Shoup-Rubin protocol, which adopts smartcards, is analysed formally. Two weaknesses due to lack of explicitness are unveiled, which affect availability to the peers of the goals of confidentiality, authentication and key distribution in our threat model.

Shoup and Rubin [148] study an existing session key distribution protocol due to Leighton and Micali [104] and prove it secure [48] using the Bellare and Rogaway's framework (2.1.4). Then, they develop a new protocol, based on the design by Leighton and Micali, for session key distribution in a three-agent setting where each agent is endowed with a smartcard that can compute a few pseudorandom functions. Finally, they extend Bellare and Rogaway's framework accounting for smartcards, and argue that the new protocol enjoys two fundamental properties. One states that a pair of agents running the protocol share the same session key at the end of a protocol session in which the Spy does not prevent the delivery of the relevant messages. There is no formal proof for this property although it may not be obvious especially if one is unfamiliar with the formalism. The other one confirms by mathematical proof that the adversary has a *negligible advantage*, signifying that the session key remains confidential. The reasoning is done without mechanised support.

We have applied the extended Inductive Method described in the previous chapter to the Shoup-Rubin protocol and verified its goals of authenticity, unicity, confidentiality, authentication and key distribution [28]. We have discovered that the confidentiality theorems that hold for the protocol model cannot be applied by the peers, so the protocol lacks goal availability (Chapter 5). This is due to the lack of explicitness in two crucial protocol steps. Inspecting the corresponding proofs suggests a simple fix, which can be verified to be effective. To our knowledge, this work represents the first mechanised proof of correctness of a full protocol based on smartcards.

This chapter presents the Shoup-Rubin protocol (§11.1), its modelling (§11.2) and its verification (§11.3). Finally, the verification is extended on an updated version of the protocol (§11.4) that achieves stronger goals.

11.1 The Shoup-Rubin Protocol

An abstract version of the protocol, obtained from both the designers and the implementors' papers, is presented in this section. An agent P's long-term key (shared with the Server) is denoted by Kp, P's smartcard by C_p and P's smartcard long-term key by K_{Cp}.

The protocol relies on the concept of *pairkey* (due to Leighton and Micali [104]) to establish a long-term secret between the smartcards of a pair of agents. The pairkey is historically associated with the pair of agents: the one for agents A and B is $\Pi_{ab} = \{A\}_{Kb} \oplus \{B\}_{Ka}$, where $\oplus$ is the bitwise exclusive-or operator. While A's card can compute $\{B\}_{Ka}$ and then $\pi_{ab} = \{A\}_{Kb}$ from Π_{ab}, B's card can compute π_{ab} directly. Hence, the two cards share the long-term secret π_{ab}, which we call *pair-k* for A and B.

$$
\begin{array}{llllll}
\text{I}: & 1. & A & \to & \text{S} & : \quad A, B \\
& 2. & \text{S} & \to & A & : \quad \Pi_{ab}, \{\Pi_{ab}, B\}_{Ka} \\[2mm]
\text{II}: & 3. & A & \to & C_a & : \quad A \\
& 4. & C_a & \to & A & : \quad Na, \{Na\}_{K_{Ca}} \\[2mm]
\text{III}: & 5. & A & \to & B & : \quad A, Na \\[2mm]
\text{IV}: & 6. & B & \to & C_b & : \quad A, Na \\
& 7. & C_b & \to & B & : \quad Nb, Kab, \{Na, Nb\}_{\pi_{ab}}, \{Nb\}_{\pi_{ab}} \\[2mm]
\text{V}: & 8. & B & \to & A & : \quad Nb, \{Na, Nb\}_{\pi_{ab}} \\[2mm]
\text{VI}: & 9. & A & \to & C_a & : \quad B, Na, Nb, \Pi_{ab}, \\
& & & & & \quad \{\Pi_{ab}, B\}_{Ka}, \{Na, Nb\}_{\pi_{ab}}, \{Na\}_{K_{Ca}} \\
& 10. & C_a & \to & A & : \quad Kab, \{Nb\}_{\pi_{ab}} \\[2mm]
\text{VII}: & 11. & A & \to & B & : \quad \{Nb\}_{\pi_{ab}}
\end{array}
$$

Fig. 11.1. Shoup-Rubin protocol

The full protocol (Figure 11.1) develops through seven phases. The odd-numbered ones take place over the network, while the even-numbered ones cover the communication between agents and smartcards.

Phase I. An initiator A tells the trusted Server that she wants to initiate a session with a responder B, and receives in return the pairkey Π_{ab} and its certificate encrypted under her long-term key.

Phase II. A queries her card and receives a fresh nonce and its certificate encrypted under the card long-term key. The form of A's query is specified neither by the designers nor by the implementors, so our choice of message 3 is arbitrary.

Phase III. A contacts B, sending him her identity and her nonce Na.

Phase IV. B queries his card with the data received from A, and obtains a new nonce Nb, the session key Kab, a certificate for Na and Nb, and a certificate for Nb; Kab is constructed as a function of Nb and π_{ab}.

Phase V. B forwards his nonce Nb and the certificate for Na and Nb to A.

Phase VI. A feeds her card B's name, the two nonces (she has just received Nb), the pairkey and its certificate, and the two certificates for the nonces; A's card computes π_{ab} from Π_{ab} and uses it with the nonce Nb to compute the session key Kab; the card outputs Kab and the certificate for Nb, which is encrypted under π_{ab}.

Phase VII. A forwards the certificate for Nb to B.

The protocol makes the assumption of secure means, so that the Spy cannot listen in between agents and their respective cards. The cards output the session keys in the clear. Although this feature may seem unrealistic to use on a vast scale, in a sense it adds robustness to a protocol by reducing each agent's knowledge to the PIN to activate his card. The current version of Shoup-Rubin in fact employs smartcards that are not PIN operated so no agent knows any long-term secrets. (The published papers never state this explicitly but Peter Honeyman, one of the implementors, kindly clarified it during a private conversation).

Nevertheless, other features may seem incautious. For example, the protocol reveals A's nonce to the Spy in step 5, and B's in step 8. An informal account of the consequences can be hardly given. We formally verify that, even if the session key is computed out of B's nonce, the knowledge of this nonce does not help the Spy discover the session key as long as she cannot use A and B's cards.

11.2 Modelling Shoup-Rubin

The protocol never uses a pairkey as a cryptographic key but merely as a means to establish the corresponding pair-k. Moreover, a pairkey remains secret as long as the Spy does not observe or compute it. Therefore, our model treats pairkeys as nonces, formalising them by the function

Pairkey : agent ∗ agent ⟶ nat

This produces a natural number to be used with message constructor Nonce. By contrast, a pair-k is used as a proper cryptographic key, and a session key is in turn constructed from a nonce and a pair-k. They are formalised respectively as

pairK : agent ∗ agent ⟶ key

sesK : nat ∗ key ⟶ key

At the operational level, we do not need to explore the implementation details beyond these components; we are interested in their abstract properties. The function Pairkey cannot be declared collision-free because it represents an application of the exclusive-or operator. As expected, this will influence the corresponding confidentiality argument. Assuming that collision of keys is impossible, the other two functions are declared as collision-free, and their ranges as disjoint. Also, they are respectively disjoint from the ranges of the functions formalising other long-term keys (§10.1.3), so that any pair-k differs from a card key and so forth.

The actual definition of initState must specify the general definition seen above (§10.3) to reflect the extra secrets involved in this protocol. Smartcards are not PIN operated here, so all occurrences of the function PIN are omitted from this presentation. The full implementation of initState, which comes with the file `Smartcard.thy` (Figure 3.1), can be found in Appendix C.1 along with a few technical lemmas.

1. The Server's initial knowledge must also comprise all pairkeys and all pair-k's.

   initState Server $\triangleq$ (Key' range crdK) $\cup$ (Key' range shrK) $\cup$
   (Key' range pairK) $\cup$ (Nonce' range Pairkey)

2. The friendly agents' initial knowledge is empty, so they are not able to reveal any secrets to the Spy.

   initState (Friend i) $\triangleq$ {}

3. Recall the definitions of pairkey and pair-k from the previous section. The Spy's initial knowledge must be extended on the pair-k for a pair of agents if the card of the second agent is cloned, because the Spy knows that agent's shared key. A pairkey must be included if both the corresponding cards are cloned.

   initState Spy $\triangleq$ (Key' crdK' cloned) $\cup$
   (Key' shrK'$\{A.$ Card $A \in$ cloned$\}) \cup$
   (Key' pairK'$\{(X, B).$ Card $B \in$ cloned$\}) \cup$
   (Key' Pairkey'$\{(A, B).$ Card $A \in$ cloned and
   Card $B \in$ cloned$\})$

The formalisations of smartcards, events and the Spy are inherited from the general treatment presented in the previous chapter. However, the Server never uses its smartcard in this protocol.

We declare the constant `sr` as a set of lists of events. It designates the formal protocol model and is defined in the rest of the section by means of inductive rules. Since the protocol assumes secure means and the cards are not PIN operated, definition 10.1.3' of illegal usability (§10.3) applies; the flag secureM must be set to true, so that the function knows remains appropriately defined. The following presentation is derived from file ShoupRubin.thy (Figure 3.1).

11.2.1 Basics

The basic rules of a formal protocol model are presented in Figure 11.2. The empty trace formalises the initial scenario, in which no protocol session has taken place. Rule *Nil* as usual sets the base of the induction stating that the empty trace is admissible in the protocol model. All other rules represent inductive steps, so they detail how to extend a given trace of the model. In particular, rule *Reception* allows messages sent on the network to be received by their respective intended recipients. Rule *Fake* is treated later (§11.2.9).

Nil:
```
[] ∈ sr
```

Reception:
```
⟦ evsR ∈ sr; Says A B X ∈ set evsR ⟧ ⟹ Gets B X # evsR ∈ sr
```

Fig. 11.2. Inductive model of Shoup-Rubin: basics

11.2.2 Phase I

The rules modelling phase I of the protocol are presented in Figure 11.3. Any agent except the Server may initiate a protocol session at any time; hence, the corresponding event may extend any trace of the model (*SR1*). The model cannot be so permissive as to dispose with condition that A is not the Server; otherwise, Theorem 11.3.1 would not hold, as clarified below (§11.3.1).

SR1:
```
⟦ evs1 ∈ sr; A ≠ Server ⟧
⟹ Says A Server {|Agent A, Agent B|} # evs1 ∈ sr
```

SR2:
```
⟦ evs2 ∈ sr; Gets Server {|Agent A, Agent B|} ∈ set evs2 ⟧
⟹ Says Server A {|Nonce (Pairkey(A,B)),
                   Crypt (shrK A) {|Nonce (Pairkey(A,B)), Agent B|}|}
   # evs2 ∈ sr
```

Fig. 11.3. Inductive model of Shoup-Rubin: phase I

Having received a message quoting two agent names — initiator and responder of the session — the Server computes the pairkey for them and sends it with a certificate to the initiator (*SR2*). Although the pairkey is sent in the clear, it does not reveal its peers. This information is carried by the certificate, which explicitly creates the association between pairkey and peers.

11.2.3 Phase II

The rules modelling phase II of the protocol are presented in Figure 11.4. The initiator of a protocol session may query her own smartcard provided that she received a message containing a nonce and a certificate (*SR3*).

SR3:
⟦ evs3 ∈ sr; legalUse(Card A);
  Says A Server ⦃Agent A, Agent B⦄ ∈ set evs3;
  Gets A ⦃Nonce Pk, Cert⦄ ∈ set evs3 ⟧
⟹ Inputs A (Card A) (Agent A) # evs3 ∈ sr

SR4:
⟦ evs4 ∈ sr; legalUse(Card A); Nonce Na ∉ used evs4; A ≠ Server;
  Inputs A (Card A) (Agent A) ∈ set evs4 ⟧
⟹ Outputs (Card A) A ⦃Nonce Na, Crypt (crdK (Card A)) (Nonce Na)⦄
    # evs4 ∈ sr

Fig. 11.4. Inductive model of Shoup-Rubin: phase II

The initiator gets no assurance that the nonce is in fact the pairkey for her and the intended responder, or that the certificate is specifically for the pairkey. Since the message traversed the network in the clear, the Spy might have tampered with it. It would seem sensible that the agent forwarded the entire message to the smartcard, which would be able to decrypt the certificate and verify the integrity and authenticity of the pairkey. However, the protocol specification does not encompass this, so we analyse the protocol with a simpler input message containing only the initiator's name. Given the input, the card issues a fresh nonce and a certificate for it (*SR4*). The card keeps no record of the nonce in order to conserve memory. By contrast, it is the certificate what will subsequently confirm the authenticity of the nonce to the card. Both steps rest on a legally usable smartcard because they express some of the legal operations by the card owner. Disposing with condition that A is not the Server would falsify a lemma that is necessary to prove the reliability Theorem 11.3.1, presented below (§11.3.1).

11.2.4 Phase III

The rules modelling phase III of the protocol are presented in Figure 11.5. When the initiator obtains a nonce and a certificate from her smartcard, she may forward the nonce along with her identity to the intended responder (*SR5*).

SR5:
⟦ evs5 ∈ sr;
  Says A Server ⦃Agent A, Agent B⦄ ∈ set evs5;
  Outputs (Card A) A ⦃Nonce Na, Cert⦄ ∈ set evs5;
  ∀ p q. Cert ≠ ⦃p, q⦄ ⟧
⟹ Says A B ⦃Agent A, Nonce Na⦄ # evs5 ∈ sr

Fig. 11.5. Inductive model of Shoup-Rubin: phase III

Later (phase V, §11.2.6), the responder obtains a message of the same
form with a different certificate, and must perform different events. At that
stage, should the responder initiate another protocol session with a third
agent, he will not be able to decide whether to behave according to phase
III or to phase V unless he checks the certificate. If it is a one-component
cipher, then phase III follows; if it is a concatenated message, then phase
V follows. These alternatives may be discerned in practice by the length of
the certificate. However, since they are mutually exclusive, our treatment of
phase III simply requires the certificate not to be a concatenated message.
Both the designers and the implementors of the protocol omit this check, and
thus introduce some ambiguity in the specification. Incidentally, it must be
recalled that, when the certificate is a cipher, no agent can check its internal
structure because its encryption key is only known to a smartcard.

11.2.5 Phase IV

The rules modelling phase IV of the protocol are presented in Figure 11.6.

SR6:
⟦ evs6 ∈ sr; legalUse(Card B);
  Gets B ⦃Agent A, Nonce Na⦄ ∈ set evs6 ⟧
⟹ Inputs B (Card B) ⦃Agent A, Nonce Na⦄ # evs6 ∈ sr

SR7:
⟦ evs7 ∈ sr; legalUse(Card B);
  Nonce Nb ∉ used evs7; Key (sesK(Nb,pairK(A,B))) ∉ used evs7;
  Inputs B (Card B) ⦃Agent A, Nonce Na⦄ ∈ set evs7⟧
⟹ Outputs (Card B) B ⦃Nonce Nb, Key (sesK(Nb,pairK(A,B))),
                      Crypt (pairK(A,B)) ⦃Nonce Na, Nonce Nb⦄,
                      Crypt (pairK(A,B)) (Nonce Nb)⦄
     # evs7 ∈ sr

Fig. 11.6. Inductive model of Shoup-Rubin: phase IV

This phase sees the responder forward a cleartext message received from
the network to his smartcard, provided that the card is legally usable (*SR6*).
The smartcard issues a fresh nonce, computes the pair-k for initiator and
responder, and uses these components to produce a session key. The nonce

being fresh, the session key is also fresh. Finally, the card outputs the nonce, the session key and two certificates (*SR7*). One certificate establishes the association between the initiator's and the responder's nonce, and will be inspected by the initiator's card in phase VI. The other certificate will be retained by the responder, who will make sure of obtaining it again from the network in the final phase.

11.2.6 Phase V

The rules modelling phase V of the protocol are presented in Figure 11.7. When the responder obtains from his card a nonce followed by a key and two certificates, he prepares for sending the nonce and one certificate to the initiator (*SR8*). However, he must recall having previously quoted the initiator's identity to the card, trusting the card output to refer to his specific input. Observe that the three components following the nonce in the card output might be seen as a unique certificate, thus inviting the ambiguity discussed above (§11.2.4).

SR8:
⟦ evs8 ∈ sr;
   Inputs B (Card B) ⦃Agent A, Nonce Na⦄ ∈ set evs8;
   Outputs (Card B) B ⦃Nonce Nb, Key K, Cert1, Cert2⦄ ∈ set evs8 ⟧
⟹ Says B A ⦃Nonce Nb, Cert1⦄ # evs8 ∈ sr

Fig. 11.7. Inductive model of Shoup-Rubin: phase V

11.2.7 Phase VI

The rules modelling phase VI of the protocol are presented in Figure 11.8. The scenario returns to the initiator. Before she queries her legally usable card, she verifies she has taken hold of three messages, each containing a nonce and a certificate. She takes on trust the nonce *Pk* as the pairkey and *Cert1* as its certificate. She recalls having obtained from her smartcard a nonce *Na* with a certificate that is not a concatenated message, which signifies that the nonce was issued for her when she was acting as initiator. Then, she treats *Nb* as the responder's nonce and *Cert3* as a certificate for *Na* and *Nb*. Finally, she feeds these components to her smartcard (*SR9*). The card checks whether all the received components have the correct form and, if so, computes the pair-k from the pairkey and then produces the session key and a certificate for the responder's nonce (*SR10*). The condition that *A* is not the Server can be justified as above with rule *SR4*.

SR9:
⟦ evs9 ∈ sr; legalUse(Card A);
  Gets A ⦃Nonce Pk, Cert1⦄ ∈ set evs9;
  Outputs (Card A) A ⦃Nonce Na, Cert2⦄ ∈ set evs9;
  Gets A ⦃Nonce Nb, Cert3⦄ ∈ set evs9;
  ∀ p q. Cert2 ≠ ⦃p, q⦄ ⟧
⟹ Inputs A (Card A) ⦃Agent B, Nonce Na, Nonce Nb, Nonce Pk,
                       Cert1, Cert3, Cert2⦄
        # evs9 ∈ sr

SR10:
⟦ evs10 ∈ sr; legalUse(Card A);   A ≠ Server;
  Inputs A (Card A) ⦃Agent B, Nonce Na, Nonce Nb,
                     Nonce (Pairkey(A,B)),
                     Crypt (shrK A) ⦃Nonce (Pairkey(A,B)), Agent B⦄,
                     Crypt (Pairkey(A,B)) ⦃Nonce Na, Nonce Nb⦄,
                     Crypt (crdK (Card A)) (Nonce Na)⦄ ∈ set evs10 ⟧
⟹ Outputs (Card A) A ⦃Key (sesK(Nb,pairK(A,B))),
                       Crypt (pairK(A,B)) (Nonce Nb)⦄
        # evs10 ∈ sr

Fig. 11.8. Inductive model of Shoup-Rubin: phase VI

11.2.8 Phase VII

The rules modelling phase VII of the protocol are presented in Figure 11.9. Upon reception of a cryptographic key and a certificate from her smartcard, the initiator forwards the certificate to the responder (*SR11*).

SR11:
⟦ evs11 ∈ sr;
  Says A Server ⦃Agent A, Agent B⦄ ∈ set evs11;
  Outputs (Card A) A ⦃Key K, Cert⦄ ∈ set evs11 ⟧
⟹ Says A B (Cert) # evs11 ∈ sr

Fig. 11.9. Inductive model of Shoup-Rubin: phase VII

11.2.9 Threats

In addition to the legal behaviour described above, the Spy may also act illegally. She observes the traffic on each trace, extracts all message components, and builds all possible fake messages to send on the network or to input to the illegally usable cards. This is modelled by rule *Fake* in Figure 11.10 (which is drawn from Figure 10.2).

We assume that the algorithm that the cards use to compute the session keys is publicly known. Therefore, should the Spy know the relevant components of a session key, she will be able to compute the key. We allow this by

Fake:
```
[ evsF ∈ sr; illegalUse(Card A);
  X ∈ synth (analz (knows Spy evsF)) ]
⟹ Says Spy B X # Inputs Spy (Card A) X # evsF ∈ sr
```

Fig. 11.10. Inductive model of Shoup-Rubin: threats on messages

Paulson's method used on the TLS protocol [134], rather than by extending the definition of synth, which would complicate the mechanisation process. If the Spy obtains a nonce and a pair-k, she can note the corresponding session key by the rule *Forge* in Figure 11.11, thus acquiring knowledge of it. Since the pair-k's are never sent on the network but merely used as encryption keys, they can only be known initially by definition of initState. This is why the third premise of the rule does not need to mention analz.

Forge:
```
[ evsFo ∈ sr; Nonce Nb ∈ analz (knows Spy evsFo);
  Key (pairK(A,B)) ∈ knows Spy evsFo ]
⟹ Notes Spy (Key (sesK(Nb,pairK(A,B)))) # evsFo ∈ sr
```

Fig. 11.11. Inductive model of Shoup-Rubin: threats on session keys

Because the means between agents and smartcards is assumed secure, the model must be extended to allow the Spy to obtain the outputs of the illegally usable cards. According to the template in Figure 10.3, we introduce a further rule for each card output. Rule *SR4_Fake* in Figure 11.12 is built from *SR4*, while analogous rules *SR7_Fake* (built from *SR7*) and *SR10_Fake* (built from *SR10*) are also needed but omitted here.

SR4_Fake:
```
[ evs4F ∈ sr; illegalUse(Card A); Nonce Na ∉ used evs4F;
  Inputs Spy (Card A) (Agent A) ∈ set evs4F ]
⟹ Outputs (Card A) Spy {|Nonce Na, Crypt (crdK (Card A)) (Nonce Na)|}
     # evs4F ∈ sr
```

Fig. 11.12. Inductive model of Shoup-Rubin: threats on card outputs

11.2.10 Accidents

The protocol model must be completed by allowing accidents (or breaches of security) on session keys, as shown in Figure 11.13. This is typically done by a single rule (as seen on BAN Kerberos, §6.4), or by two rules leaking two different kinds of session keys (as seen on Kerberos IV, §7.2.5).

OopsB:
⟦ evsOb ∈ sr;
  Outputs (Card B) B ⦃Nonce Nb, Key K, Cert,
                       Crypt (pairK(A,B)) (Nonce Nb)⦄
    ∈ set evsOb ⟧
⟹ Notes Spy ⦃Key K, Nonce Nb, Agent A, Agent B⦄ # evsOb ∈ sr

OopsA:
⟦ evsOa ∈ sr;
  Outputs (Card A) A ⦃Key K, Crypt (pairK(A,B)) (Nonce Nb)⦄
    ∈ set evsOa ⟧
⟹ Notes Spy ⦃Key K, Nonce Nb, Agent A, Agent B⦄ # evsOa ∈ sr

Fig. 11.13. Inductive model of Shoup-Rubin: accidents

Shoup-Rubin requires both peers to handle the same session key, respectively in phases IV and VI. Therefore, the Spy has a chance to discover the session key from both of them. In the worst case, she will also discover the nonce used to compute the key and the identity of its peers (*OopsA* and *OopsB*).

The Spy cannot learn any pair-k's by accident because no agent ever sees any. By definition of initState, she can only know some initially by exploiting the relevant cloned cards.

11.3 Verifying Shoup-Rubin

In general, it may be useful to interpret the guarantees proved for a smart-card protocol also from the viewpoint of smartcards, possibly helping optimise their hardware or software design. This section discusses those established about Shoup-Rubin; hence, *evs* is a generic trace of the formal protocol model sr. Observe that a guarantee that requires inspecting the form of a certificate may be useful to cards but never to agents, who cannot decipher any certificates since they know no long-term keys. The minimal trust now often includes that certain cards not be usable by the Spy.

The reliability theorems show that the model makes the expected use of smartcards (§11.3.1) and that messages 7 and 10 crucially lack some explicitness. Suitable regularity lemmas can be expressed about all three kinds of long-term keys employed by the protocol (§11.3.2). While the authenticity argument (§11.3.3) only yields a single guarantee for the card that belongs to the protocol initiator, the unicity argument (§11.3.4) will provide guarantees for both initiator and responder. Confidentiality (§11.3.5) is weakened by the mentioned lack of explicitness, as are the goals of authentication (§11.3.6) and key distribution (§11.3.7). Theorem names follow our usual conventions (§1.3.2).

11.3.1 Reliability of the Shoup-Rubin Model

The model Server functions reliably (Theorem 11.3.1). However, this theorem cannot be made useful to A (the way Theorem 7.3.1 was by Theorem 7.3.3). The authenticity argument about the message ⦃Nonce Pk, $Cert$⦄ is extremely weak. Should A receive such a message, she cannot be guaranteed that it is an instance of message 2, namely that the Server sent it, because the message is concatenated; nor can she inspect the form of the certificate.

Theorem 11.3.1 (SR_Says_Server_message_form). *If evs contains*

   Says Server A ⦃Nonce Pk, $Cert$⦄

then, for some B,

   $Pk = $ Pairkey(A, B) *and*
   $Cert = $ Crypt(shrK A)⦃Nonce (Pairkey(A, B)), Agent B⦄.

As mentioned above, the proof relies on the assumption that A is not the Server made in rule *SR1*, which would introduce an event falsifying the conclusion of the theorem. It also relies on a subsidiary lemma stating that the Server never uses his smartcard (**SR_Outpts_Server_not_evs**, omitted here); otherwise, rules *SR8* or *SR11* would also introduce events falsifying the theorem. In turn, that lemma only holds if the condition that A is not the Server is added to rule *SR4*.

   Further guarantees concern the use of the smartcards allowed by the protocol, the outputs that they produce and the inputs that uncompromised agents send them.

On the use of the smartcards. If an agent other than the Spy queries a smartcard or receives a message from it, then the card must belong to that agent and must be legally usable (Theorem 11.3.2). Hence, that agent can only use his own card and can only use it legally, as we required.

Theorem 11.3.2 (SR_Inputs_Outpts_Card). *If A is not the Spy and evs contains either*

   Inputs $A\,C\,X$ *or*   Outpts $C\,A\,Y$

then

   $C = $ Card A *and*   legalUse(Card A).

   Our Spy can act both legally and illegally. In fact, if the Spy uses a smartcard, then the card must be either the Spy's own card, which is legally usable, or some other agent's card that is illegally usable (Theorem 11.3.3). Since the Spy's card is not illegally usable, the agent A mentioned by the theorem certainly differs from the Spy.

Theorem 11.3.3 (SR_Inputs_Card_Spy). *If evs contains either*

   Inputs Spy $C\,X$ *or*   Outpts Spy $A\,Y$

then

$(C = \mathsf{Card\,Spy}$ *and* $\mathsf{legalUse}(\mathsf{Card\,Spy}))$ *or*
$(\exists A.\ C = \mathsf{Card}\,A$ *and* $\mathsf{illegalUse}(\mathsf{Card}\,A)).$

On the outputs of the smartcards. To establish that the model smart-cards work reliably, two categories of guarantees can be proved for the Outputs events.

One category states that the cards only give the correct outputs when fed the expected inputs, so the cards cannot grant the Spy unlimited resources. The case for step 10 of the protocol is presented below (Theorem 11.3.4), while those for steps 4 and 7 are similar and omitted here.

Theorem 11.3.4 (SR_Outpts_which_Card_10). *If evs contains*

$\mathsf{Outputs}\,(\mathsf{Card}\,A)\ A\ \{\!|\mathsf{Key}\,(\mathsf{sesK}(Nb,\mathsf{pairK}(A,B))),$
$\qquad\qquad\qquad\quad \mathsf{Crypt}(\mathsf{pairK}(A,B))(\mathsf{Nonce}\,Nb)|\!\}$

then, for some Na, evs also contains

$\mathsf{Inputs}\ A\ (\mathsf{Card}\,A)\ \{\!|\,\mathsf{Agent}\,B,\mathsf{Nonce}\,Na,\mathsf{Nonce}\,Nb,\mathsf{Nonce}\,(\mathsf{Pairkey}(A,B)),$
$\qquad\qquad\quad \mathsf{Crypt}(\mathsf{shrK}\,A)\{\!|\mathsf{Nonce}\,(\mathsf{Pairkey}(A,B)),\mathsf{Agent}\,B|\!\},$
$\qquad\qquad\quad \mathsf{Crypt}(\mathsf{pairK}(A,B))\{\!|\mathsf{Nonce}\,Na,\mathsf{Nonce}\,Nb|\!\},$
$\qquad\qquad\quad \mathsf{Crypt}(\mathsf{crdK}(\mathsf{Card}\,A))(\mathsf{Nonce}\,Na)\,|\!\}.$

Another category of reliability theorems confirms that the card CPUs function correctly. Therefore, given a specific output, the form of its com-ponents can be tracked down. One such guarantee can be established on an instance of message 4 (Theorem 11.3.5). The length of the certificate must be checked because of the protocol ambiguity already encountered (§11.2.4). Recall that an event $\mathsf{Outputs}\,C\,A\,X$ also models A's reception of X, so the theorem is applicable also by A.

Theorem 11.3.5 (SR_Outpts_A_Card_form_4). *If evs contains*

$\mathsf{Outputs}\,(\mathsf{Card}\,A)\ A\ \{\!|\mathsf{Nonce}\,Na,Cert|\!\}$

and Cert is not concatenated, then

$Cert = \mathsf{Crypt}(\mathsf{crdK}(\mathsf{Card}\,A))(\mathsf{Nonce}\,Na).$

Analogous considerations apply to message 7. Upon B's reception of an output, we can guarantee its form for some peer A and some nonce Na (The-orem 11.3.6). The existential form of the assertion says that B receives the session key in a message that does not inform him of the peer with whom the key is to be used. This violates a well-known explicitness principle, per-haps unknown at the time of the design, due to Abadi and Needham: "Every message should say what it means. The interpretation of the message should depend only on its content" [7, §2.1]. The underlying transport protocol can-not reveal the peer's identity either. If B uses the session key with the wrong peer, the consequences should not be disastrous provided that the key remains confidential. But, this lack of explicitness does weaken the confidentiality, authentication and key distribution guarantees accomplished by the protocol, as discussed below.

Theorem 11.3.6 (SR_Outpts_B_Card_form_7). *If evs contains*

Outputs (Card B) B {Nonce Nb, Key Kab, $Cert1$, $Cert2$}

then, for some A and Na,

$Kab = \mathsf{sesK}(Nb, \mathsf{pairK}(A, B))$ *and*
$Cert1 = \mathsf{Crypt}(\mathsf{pairK}(A, B))\{$Nonce Na, Nonce $Nb\}$ *and*
$Cert2 = \mathsf{Crypt}(\mathsf{pairK}(A, B))($Nonce $Nb)$.

The card CPUs are also reliable when producing an instance of message 10 (Theorem 11.3.7). The existential form of the assertion reveals another lack of explicitness in the protocol design: the identity of the peer with whom to use the key is not specified. When A receives the session key, she has to guess the peer who shares it. This task is entirely heuristic in our threat model, as the card might give its outputs in an unspecified order. Similarly, the message fails to mention the nonce associated with the session key.

Theorem 11.3.7 (SR_Outpts_A_Card_form_10). *If evs contains*

Outputs (Card A) A {Key Kab, $Cert$}

then, for some B and Nb,

$Kab = \mathsf{sesK}(Nb, \mathsf{pairK}(A, B))$ *and*
$Cert = \mathsf{Crypt}(\mathsf{pairK}(A, B))($Nonce $Nb)$.

The theorem also shows that step 10 binds the form of the session key to the card that creates it, and associates the session key with the certificate. Therefore, should the former be inspectable, the structure of the latter could be derived (Theorem 11.3.8), and vice versa.

Theorem 11.3.8 (SR_Outpts_A_Card_form_10_bis). *If evs contains*

Outputs (Card A) A {Key $(\mathsf{sesK}(Nb, \mathsf{pairK}(A', B)))$, $Cert$}

then,

$A = A'$ *and* $Cert = \mathsf{Crypt}(\mathsf{pairK}(A, B))($Nonce $Nb)$.

On the inputs of the smartcards. Analogous categories of guarantees can be established for the Inputs events.

Agents other than the Spy must use the legally usable smartcards in a legal manner. Therefore, they produce inputs whose origin can be documented. For example, let us assume that an agent A queries a card as in step 9 of the protocol (by Theorem 11.3.4, the card belongs to A) quoting an agent B. We can prove that A initiated a session with B, and received the components of the query either from the network or from the card, by means of suitable events (Theorem 11.3.9).

Theorem 11.3.9 (SR_Inputs_A_Card_9). *If A is not the Spy and evs contains*

Inputs $A\,C$ {|Agent B, Nonce Na, Nonce Nb, Nonce Pk,
$\qquad$ *Cert1*, *Cert2*, *Cert3*|}

then evs also contains

Says A Server {|Agent A, Agent B|}   *and*
Gets A {|Nonce Pk, *Cert1*|}   *and*
Gets A {|Nonce Nb, *Cert2*|}   *and*
Outputs $C\,A$ {|Nonce Na, *Cert3*|}.

Although the first event of the conclusion highlights A's intention to communicate with B, none of the remaining events mentions B. So, A cannot be assured she is feeding her card the components meant for the session with B. Even if the Gets events mentioned B, his identity would not be reliable as the Spy can tamper with concatenated messages coming from the network. The Outputs event could mention B reliably as it takes place over a secure means, but fails to do so. However, by Theorem 11.3.4, A will get an output from her card only if she uses the correct components as input.

Similar theorems pertain to the other queries to the smartcards, steps 3 and 6 of the protocol.

The form of the inputs created in steps 3 and 6 of the protocol are self-explanatory. Step 9 is more complicated. While most of such input was exposed to the network risks, the certificate produced earlier by A's card was not, so its form can be derived (Theorem 11.3.10) signifying that all agents use their own card correctly.

Theorem 11.3.10 (SR_Inputs_A_Card_form_9). *If evs contains*

Inputs A (Card A) {|Agent B, Nonce Na, Nonce Nb, Nonce Pk,
$\qquad$ *Cert1*, *Cert2*, *Cert3*|}

then

$Cert3 = \mathsf{Crypt}(\mathsf{crdK}(\mathsf{Card}\,A))(\mathsf{Nonce}\,Na).$

Observe that the guarantee also applies to the Spy's use of her own card. Upon reception of a message, A's card can determine whether it is an instance of message 9 by looking at its cleartext part. The card should inspect carefully the second and third certificates because they could be fake. Their form is in fact not provable in the model. Having proved the integrity of the third certificate may suggest that it is superfluous to the design, and that the card can avoid checking it. Nevertheless, should an agent insert a fake nonce as second component of message 9, inspection of the third certificate would detect the misbehaviour. However, any agent other than the Spy only acts legally.

11.3.2 Regularity

The protocol sends no long-term keys over the network, so the Spy could do so if and only if she knows them before the protocol begins. The Spy can discover a card key and the card owner's key only from cloning the card (see definition of initState, §11.2). Using the latter key, she can compute all the pair-k's meant for the card owner. The three relevant regularity lemmas follow.

Lemma 11.3.1 (SR_Spy_analz_shrK). *Trace evs is such that* Key (shrK A) $\in$ analz(knows Spy *evs*) *if and only if* Card A $\in$ cloned.

Lemma 11.3.2 (SR_Spy_analz_crdK). *Trace evs is such that* Key (crdK C) $\in$ analz(knows Spy *evs*) *if and only if* C $\in$ cloned.

Lemma 11.3.3 (SR_Spy_analz_pairK). *Trace evs is such that* Key (pairK(P, B)) $\in$ analz(knows Spy *evs*) *if and only if* Card B $\in$ cloned.

11.3.3 Authenticity

All Shoup-Rubin's certificates are sealed with long-term keys, so the agents do not directly get authenticity guarantees about them. However, since the long-term keys are stored into the smartcards, the authenticity argument in general can be formulated for the smartcards, possibly helping to optimise their design. The agents may get authenticity guarantees indirectly from the smartcards. Shoup-Rubin inputs a smartcard with encrypted certificates only in step 9. We develop the corresponding authenticity argument via some subsidiary authenticity lemmas that are not directly applicable by either agents or cards. Incidentally, due to the assumption of secure means, if a card receives a certificate as part of an input in a trace *evs*, it cannot be immediately concluded by definition of knows (§10.3) that the certificate is in the network traffic, namely in parts(knows Spy *evs*). But perhaps it can be concluded by other methods due to specific features of the protocol.

Along with a pairkey, the Server issues a certificate that verifies it. When the certificate is in the traffic, we can prove that it originated with the Server if the regularity Lemma 11.3.1 is applicable. Therefore, given that the peer's card is not cloned, the certificate is authentic (Lemma 11.3.4). At this stage, the form of the pairkey can be specified via Theorem 11.3.1.

Lemma 11.3.4 (SR_Pairkey_certificate_authentic). *If A's card is not cloned and evs is such that*

Crypt(shrK A){Nonce Pk, Agent B} $\in$ parts(knows Spy *evs*)

then evs contains

Says Server A {| Nonce Pk, Crypt(shrK A){| Nonce Pk, Agent B |} |}

and

$Pk = \mathsf{Pairkey}(A, B)$.

We can verify formally that the certificate that associates A and B's nonces is built in step 7 (Lemma 11.3.5). Since the certificate is sealed with the corresponding pair-k, investigating its origin requires an appeal to the regularity Lemma 11.3.3, which prescribes B's card not to be cloned. However, a stronger assumption is needed on B's card to let us solve case *SR7_Fake*: the card must not be illegally usable; otherwise, it could also output towards the Spy.

Lemma 11.3.5 (SR_Na_Nb_certificate_authentic). *If B's card is not illegally usable and evs is such that*

Crypt(pairK(A, B)){| Nonce Na, Nonce Nb |} $\in$ parts(knows Spy evs)

then evs contains

Outputs (Card B) B {| Nonce Nb, Key (sesK(Nb, pairK(A, B)))),
                Crypt(pairK(A, B)){| Nonce Na, Nonce Nb |},
                Crypt(pairK(A, B))(Nonce Nb) |}.

Message 7 ends with another certificate that verifies B's nonce, namely $\{ Nb \}_{\pi_{ab}}$. So, we can prove a theorem, omitted here, that is identical to Theorem 11.3.5, except for the certificate considered and the assertion now being in the scope of an existentially quantified nonce Na (**SR_Nb_certificate_authentic_bis**, omitted here). The same certificate is also output by A's card in message 10. Proving this result (Lemma 11.3.6) also requires B not to be the Spy in order to solve case *SR7* (so that the corresponding event does not introduce the certificate in the traffic), and A's card not to be illegally usable to solve case *SR10_Fake*.

Lemma 11.3.6 (SR_Nb_certificate_authentic). *If B is not the Spy, A and B's cards are not illegally usable and evs is such that*

Crypt(pairK(A, B))(Nonce Nb) $\in$ parts(knows Spy evs)

then evs contains

Outputs (Card A) A {| Key (sesK(Nb, pairK(A, B)))),
                Crypt(pairK(A, B))(Nonce Nb) |}.

The authenticity lemmas serve to prove an authenticity theorem that is applicable by A's card (Theorem 11.3.11). So, the theorem must include the assumptions on agents and cards required by the lemmas. Upon reception of message 9, the card must inspect the first two certificates, as advised by

Theorem 11.3.10. If the first certificate has the expected form, then Theorem 11.3.9 and Lemma 11.3.4 (plus the basic lemma that once a message is received, its components appear in the traffic) prove the first event of the assertion. Hence the assumptions that A is not the Spy and that her card is not cloned. These cannot be replaced by the assumption of A's card not being illegally usable, which is weaker. Similarly, if the second certificate is as expected, then Theorem 11.3.9 and Lemma 11.3.5 prove the second event. Hence the assumption that B's card is not illegally usable. The third certificate does not need to be inspected thanks to its provable integrity, so Theorem 11.3.9 alone justifies the third event of the assertion. This reasoning is mechanisable by one Isabelle command that applies Theorem 11.3.9 only once, and then the necessary authenticity lemma.

Theorem 11.3.11 (SR_Inputs_A_Card_9_authentic). *If A is not the Spy, A's card is not cloned, B's card is not illegally usable and evs contains*

> Inputs A (Card A) $\{\!|$ Agent B, Nonce Na, Nonce Nb, Nonce Pk,
> Crypt(shrK A)$\{\!|$Nonce Pk, Agent $B\}\!|$,
> Crypt(pairK(A, B))$\{\!|$Nonce Na, Nonce $Nb\}\!|$, $Cert3$ $\}\!|$

then evs also contains

> Says Server A $\{\!|$Nonce Pk, Crypt(shrK A)$\{\!|$Nonce Pk, Agent $B\}\!|\}\!|$ *and*

> Outputs (Card B) B $\{\!|$ Nonce Nb, Key (sesK$(Nb,$ pairK$(A, B))$)),
> Crypt(pairK(A, B))$\{\!|$Nonce Na, Nonce $Nb\}\!|$,
> Crypt(pairK(A, B))(Nonce Nb) $\}\!|$ *and*

> Outputs (Card A) A $\{\!|$Nonce Na, $Cert3\}\!|$.

The authenticity of the crucial message components can be investigated in the same fashion as that of the certificates. Let us consider the authenticity of pairkeys. Only the Server is entitled to issue Pairkey(A, B), which does not belong to the initial knowledge of the Spy if either A or B's card is not cloned. This may let us believe that, if for example A's card is not cloned and that pairkey is in the traffic, then the Server must have issued it. This conjecture is easy to formalise (Conjecture 11.3.1).

Conjecture 11.3.1 (SR_Pairkey_authentic). If A's card is not cloned and evs is such that

> Pairkey(A, B) $\in$ parts(knows Spy evs)

then, for some $Cert$, evs contains

> Says Server A $\{\!|$Nonce (Pairkey(A, B)), $Cert\}\!|$.

A proof attempt leaves us, in particular, with the subgoal in Figure 11.14, which arises from case *Nil*. It cannot be derived that $A = A'$ because the pairkey is implemented in terms of the exclusive-or operator, which is not collision-free. The subgoal can be in fact falsified because there may exist two pairs of distinct agents A, A' and B, B' who satisfy the premises. In consequence, the conjecture does not hold. This proof attempt teaches us that the Spy might exploit the collisions suffered by the exclusive-or operator and forge a pairkey without knowing its original components. The probability of this happening is influenced by the redundancy introduced by the encryption function and by the length of the ciphers.

```
⟦ Card A ∉ cloned;
  Pairkey(A,B) = Pairkey (A',B');
  Card A' ∈ cloned; Card B' ∈ cloned ⟧ ⟹ False
```

Fig. 11.14. Proving pairkey authenticity for Shoup-Rubin: failed

We now examine the authenticity of the session key. This crucial message component is only sent between cards and agents, never over the network. Despite this, the Spy could either forge it (by *Forge*), or obtain it from her own card if she is one of the peers (by *SR7* or *SR10*), or learn it from the illegally usable cards (by *SR7_Fake* or *SR10_Fake*). Let us make the assumptions that prevent all these circumstances. For example, if the responder's card is not illegally usable and therefore not cloned, then the session key cannot be forged by Lemma 11.3.3. Then, if a session key ever appears in the traffic, one of its peers necessarily leaked it by accident, while the trace recorded the corresponding oops event (Lemma 11.3.7). This turns out to be a counterguarantee of authenticity because it emphasises the conditions under which a session key that is in the traffic is not authentic: the Spy in fact introduced it. However, it will be fundamental to assess a form of session key confidentiality.

Lemma 11.3.7 (SR_sesK_authentic). *If A and B are not the Spy, their cards are not illegally usable and evs is such that*

Key $(\mathsf{sesK}(Nb, \mathsf{pairK}(A, B))) \in \mathsf{parts}(\mathsf{knows}\,\mathsf{Spy}\ evs)$

then evs contains

Notes Spy $\{$Key $(\mathsf{sesK}(Nb, \mathsf{pairK}(A, B))),$ Nonce $Nb,$ Agent $A,$ Agent $B\}$.

Proving the authenticity lemmas requires a common method (simpler than Paulson's for the authenticity theorems on traditional protocols [133, §4.7]). We present below the method for Shoup-Rubin, which can be generalised straightforwardly to any smartcard protocol.

1. Apply induction.
2. If the lemma concerns
    – a certificate sealed with a shared key, then simplify case *Fake* by Lemma 11.3.1;
    – a certificate sealed with a card key, then simplify case *Fake* by Lemma 11.3.2;

  - a certificate sealed with a pair-k, then simplify case *Fake* by Lemma
    11.3.3;
  - a session key, then apply "$H \subseteq \text{parts}\,H$" to case *Forge* and simplify it
    by Lemma 11.3.3.
3. Solve case *Fake* by the standard method spy_analz [133, §4.5].
4. Apply the theorems confirming that the cards function reliably as follows:
   Theorem 11.3.5 to case *SR9*, Theorem 11.3.6 to cases *SR8* and *OopsB*,
   Theorem 11.3.7 to case *SR11*, and a variant of Theorem 11.3.7 — which
   binds the form of the certificate, given the form of the session key — to
   case *OopsA*.
5. Simplify remaining cases.

11.3.4 Unicity

Shoup-Rubin requires B's card to build a fresh session key in message 7. The
key is bound uniquely to the remaining components of the message (Theorem 11.3.12). In proving this result, after induction and simplification two
subgoals remain, which are about *SR7* and *SR7_Fake*. The latter is easily solv-
able because it forces the Spy to use her own card illegally, which is impossible. The other case is solved by freshness: the session key could not appear before. Message 7 also contains B's fresh nonce, so a variant of the theorem may
be proved using the nonce as a pivot (**SR_Outpts_B_Card_unique_nonce**,
omitted here).

Theorem 11.3.12 (SR_Outpts_B_Card_unique_key). *If evs contains*

> $\text{Outputs}\,(\text{Card}\,B)\,B\,\{\!|\text{Nonce}\,Nb,\text{Key}\,Kab,\,Cert1,\,Cert2|\!\}$ *and*
>
> $\text{Outputs}\,(\text{Card}\,B')\,B'\,\{\!|\text{Nonce}\,Nb',\text{Key}\,Kab,\,Cert1',\,Cert2'|\!\}$

then

> $B = B'$ *and* $Nb = Nb'$ *and* $Cert1 = Cert1'$ *and* $Cert2 = Cert2'$.

A similar theorem (**SR_Outpts_A_Card_unique_nonce**, omitted here)
holds for the output of step 4, exploiting the freshness of A's nonce. More
surprisingly, it holds for the output of message 10 too (Theorem 11.3.13);
Theorem 11.3.8 supplies to the fact that the card uses no fresh components.
Whenever a specific session key appears, the form of the corresponding certificate can be assessed, so the same key cannot stand by two different certificates. This method solves the subgoal about *SR10*, while the one about
SR10_Fake is terminated routinely.

Theorem 11.3.13 (SR_Outpts_A_Card_unique_key). *If evs contains*

> $\text{Outputs}\,(\text{Card}\,A)\,A\,\{\!|\text{Key}\,Kab,\,Cert|\!\}$ *and*
>
> $\text{Outputs}\,(\text{Card}\,A')\,A'\,\{\!|\text{Key}\,Kab,\,Cert'|\!\}$

then

$A = A'$ and $Cert = Cert'$.

The unicity theorems may teach agents a lot. For example, if in the real world B receives the same session key within two different instances of message 7, he may suspect that something wrong has happened. Having violated Theorem 11.3.12, the scenario is due to problems that lie outside our model, ranging from a weird malfunction of B's card to a Spy's violation of the assumption of secure means. Theorem 11.3.13 provides the equivalent guarantee to A.

However, if B happens to receive the same session key within the same message more than once, Theorem 11.3.12 will not be violated. Still, the scenario is unaccountable in the model, and thus should alarm B. Since all card CPUs function correctly, B's card must always compute a fresh key. A further guarantee may be designed to assist B in this circumstance. Upon reception of any output commencing with a nonce, B can be assured that the corresponding event is unique (Theorem 11.3.14). After expanding the definition of the predicate, the cases about $SR4$ and $SR7$ are solved by freshness of the nonce. The result also applies to the output of A's card in step 4. No similar theorem can be established about step 10, which does not involve any fresh components.

Theorem 11.3.14 (SR_Outpts_A_Card_Unique). *If evs contains*

Outputs (Card B) B {|Nonce Nb, $rest$|}

then

Unique (Outputs (Card B) B {|Nonce Nb, $rest$|}) on *evs*.

11.3.5 Confidentiality

Some counterguarantees of confidentiality can be easily obtained. Even if a specific pairkey has not been issued by the Server and its components cannot be forged, the pairkey cannot be proved confidential because of the weakness discovered via the authenticity argument (§11.3.3). Also, it is straightforward to observe from messages 6 and 8 that neither A's nor B's nonces remain confidential.

On the contrary, the regularity lemmas may be viewed as non-trivial confidentiality guarantees. Moreover, applying analz parts H to the authenticity Lemma 11.3.7, we obtain a guarantee of session key confidentiality. Any session key that cannot be forged and that has not been leaked by accident is confidential (Theorem 11.3.15). Unfortunately, the theorem is not useful to agents because the structure of the session key must be inspected.

Theorem 11.3.15 (SR_Confidentiality). *If A and B are not the Spy, their cards are not illegally usable and evs does not contain*

Notes Spy {|Key (sesK(Nb, pairK(A, B))), Nonce Nb, Agent A, Agent B|}

then evs is such that

$\mathsf{Key}\,(\mathsf{sesK}(Nb, \mathsf{pairK}(A, B))) \notin \mathsf{analz}(\mathsf{knows\,Spy}\;evs).$

This result cannot be strengthened sufficiently: we have discovered that the theorems of session key confidentiality cannot be applied by the peers within their respective minimal trust due to the lack of explicitness that affects two protocol steps [25]. It follows that the Shoup-Rubin protocol grants weak confidentiality guarantees to its peers unless the design is slightly modified (§11.4). However, the guarantees presented below can be applied by the smartcards, which might be a significant outcome for a smartcard protocol.

Paulson's general method for verifying confidentiality (§4.5) can be followed here. The necessary simplification law for analz, the session key compromise theorem, is fairly easy to obtain, since Shoup-Rubin never sends session keys over the network. The confidentiality argument for a protocol responder B must be developed on the basis of an event that B can verify, namely that his card sends a message that contains the session key. This takes place in step 7 of the protocol, which is formalised by the event

$\mathsf{Outputs}\,(\mathsf{Card}\,B)\,B\,\{\!|\mathsf{Nonce}\,Nb, \mathsf{Key}\,Kab, Cert1, Cert2|\!\}$

This includes two certificates, *Cert1* and *Cert2*, that B cannot inspect because they are sealed with specific long-term keys (no agent knows any).

We have attempted to prove *Kab* confidential in a trace *evs* that contains no oops event leaking *Kab* but that does contain the mentioned event. Also, B's card must be assumed not to be cloned, or the Spy would otherwise know pairK(P, B) for any agent P, in which case she would be able to forge the session key by rule *Forge*. The proof leaves two subgoals unsolved, respectively arising from cases *SR10* and *SR10_Fake*. The inspection of the former teaches us that B's peer might be the Spy, who could obtain a copy of *Kab* from her own smartcard. The latter subgoal shows that B's peer's card could be illegally usable regardless of the identity of the peer; the Spy would be able to use this card to compute *Kab*. While the protocol requires B's card to issue a new session key in step 7, his peer in fact computes a copy of the key from available components in step 10.

Therefore, further assumptions are necessary on B's peer and her card, but the message output to B does not state the identity of such a peer. This signifies that B does not obtain explicit information about the peer with which the session key is to be used, which again violates the explicitness principle of Abadi and Needham: "If the identity of a principal is essential to the meaning of a message, it is prudent to mention the principal's name explicitly in the message" [7, §4]. If either one of the certificates is inspected, then B's peer, A, becomes explicit, so the relevant assumptions can be stated and Theorem 11.3.16 proved.

Theorem 11.3.16 (SR_Confidentiality_B). *If A and B are not the Spy, A's card is not illegally usable, B's card is not cloned and evs contains*

Outputs (Card B) B {|Nonce Nb, Key Kab, $Cert$,
Crypt(pairK(A, B))(Nonce Nb)|}

but does not contain

Notes Spy {|Key Kab, Nonce Nb, Agent A, Agent B|}

then evs is such that

Key $Kab \notin$ analz(knows Spy evs).

From B's viewpoint, trusting that the peer is not malicious and her card cannot be used by the malicious entity is indispensable. So is trusting that the key has not been leaked by accident. These assumptions belong to B's minimal trust. What is more important is that B cannot verify that the main event of the theorem ever occurs because he cannot inspect the certificate. Therefore, he cannot apply the theorem. Our conclusion is that the protocol fails to make session key confidentiality available to B in our threat model.

Shoup and Rubin's analysis based on provable security asserts an analogous property requiring that the peers' cards be *unopened* [148, §3.1], which may be interpreted as *not cloned* in our treatment. However, their analysis does not report lack of explicitness, whereas ours does, thanks to the use of goal availability. In fact, from B's viewpoint, trusting his peer and his peer's card belongs to his own minimal trust, but trusting every agent does not. The protocol does not allow B to learn which peer to trust in the given threat model.

Similar considerations arise when reasoning from A's viewpoint. The attempt to prove confidentiality on the assumption that the event

Outputs (Card A) A {|Key Kab, $Cert$|}

formalising step 10 occurs, leaves the subgoals arising from *SR7* and *SR7_Fake* unsolved. They highlight that A could be communicating either with the Spy or with an agent whose card is illegally usable. In fact, step 10 fails to express A's peer. Also this theorem, as the previous one, can be proved if the form of *Cert* is explicit, resulting in a guarantee that can be applied by A's card but not by A (Theorem 11.3.17).

Theorem 11.3.17 (SR_Confidentiality_A). *If A and B are not the Spy, A and B's cards are not illegally usable and evs contains*

Outputs (Card A) A {|Key Kab, Crypt(pairK(A, B))(Nonce Nb)|}

but does not contain

Notes Spy {|Key Kab, Nonce Nb, Agent A, Agent B|}

then evs is such that

Key $Kab \notin$ analz(knows Spy evs).

Our conclusion is that the protocol fails to make session key confidentiality available to A in our threat model. However, Theorems 11.3.16 and 11.3.17 must be interpreted with care, as this type of protocol should also provide guarantees to the smartcards. It can be seen that these guarantees are applicable by the smartcards, which can check their main assumptions. The lack of explicitness only exists for the agents.

11.3.6 Authentication

The lack of explicitness that affects messages 7 and 10 also weakens the goals of authentication. Only A's card obtains a useful guarantee and so can detect whether certain components are being used with the wrong peer. The proof script pertaining to the following guarantees can be found in Appendix C.3.

Phase V terminates B's role in the protocol. Then, B's peer, A, obtains the session key from message 10, but the identity of B remains unspecified unless the certificate is inspected. If the certificate is not fake, then it must have originated with the instance of message 7 that concerns B (Theorem 11.3.18). The proof observes that the event of the assumption implies that the certificate $\{Na, Nb\}_{\pi_{ab}}$ appears in the traffic for some Na; then, it applies Lemma 11.3.5. This theorem is not applicable by A, who cannot check its main assumption, hence the corresponding goal is not available to her. But it is significant to A's card, which can inspect the certificate. When the card issues A with the session key, it is guaranteed that both B and his card were present on the network and that B's card, which is using the pair-k for A and B, is participating in a session with A.

Theorem 11.3.18 (SR_A_authenticates_B). *If B's card is not illegally usable and evs contains*

Outputs $(\text{Card } A)\, A\, \{\!|\, \text{Key } Kab, \text{Crypt}(\text{pairK}(A, B))(\text{Nonce } Nb)\,|\!\}$

then, for some Na, evs also contains

Outputs $(\text{Card } B)\, B\, \{\!|\, \text{Nonce } Nb, \text{Key } Kab,$
$\qquad\qquad \text{Crypt}(\text{pairK}(A, B))\{\!|\text{Nonce } Na, \text{Nonce } Nb|\!\},$
$\qquad\qquad \text{Crypt}(\text{pairK}(A, B))(\text{Nonce } Nb)\,|\!\}.$

This result may be interpreted as weak agreement of B's card with A's. The cards know the components of their outputs, so the result may also be viewed as non-injective agreement of B's card with A's on Kab. However, expressing this formally requires extending the function knows on smartcards (§10.3), which seems questionable because of the limited memory of the cards.

A variant of the theorem just given can be proved by replacing its main assumption with an earlier event: A's reception of an instance of message 8 (**SR_A_authenticates_B_Gets**, omitted here). However, our conclusions do not change because the variant requires inspection of the certificate that arrives with the message, which A is not able to do.

A relevant authentication guarantee for B should establish that A is active after B creates the session key. At the end of the protocol, B may receive from the network the certificate for his nonce. Provided that Lemma 11.3.6 is applicable, A's card can be proved to have sent a suitable instance of message 10, which establishes the presence of A and her card, and A's card intention to communicate with B (Theorem 11.3.19).

Theorem 11.3.19 (SR_B_authenticates_A). *If B is not the Spy, A and B's cards are not illegally usable and evs contains*

Gets B (Crypt(pairK(A, B))(Nonce Nb))

then evs also contains

Outputs (Card A) A { Key (sesK(Nb, pairK(A, B))),
    Crypt(pairK(A, B))(Nonce Nb) }.

Is this theorem useful to B? The answer is "no" because the agent cannot inspect the encrypted certificate. So, in practice B obtains no information about the sender of the certificate, and his peer remains unknown. Observing that the certificate was originally created in message 7 does not help because neither that message states the peer (Theorem 11.3.6). A possible method, which we verified, to make the authentication goal available to B is to conclude the protocol with two additional steps: B forwards the certificate to his card, and the card responds with A's identity. The card should use the right pair-k to decrypt the certificate, thus identifying A. While adding explicitness to message 7 is a simpler fix (as demonstrated below, §11.4), making the guarantee available also to B's card necessarily requires the additional steps.

11.3.7 Key Distribution

The key distribution guarantees discussed here are the strongest that can be proved, but are not applicable by the peers. Hence, we conclude that the protocol fails to make key distribution available. However, we shall see that one guarantee is applicable by A's card.

Applying the definition of knows to the conclusion of Theorem 11.3.18 indicates that, when A's card computes the session key for A, the key is already known to B (Theorem 11.3.20). But A cannot profit from this result. The theorem does not prevent B from being the Spy.

Theorem 11.3.20 (SR_A_keydist_to_B). *If B's card is not illegally usable and evs contains*

Outputs (Card A) A {Key Kab, Crypt(pairK(A, B))(Nonce Nb)}

then evs is such that

Key $Kab \in$ analz(knows B evs).

Let us attempt to design the corresponding guarantee for B. His session key is obtained via message 7. By Theorem 11.3.19, if B receives the last message of the protocol, he infers that A obtained *some* session key. The two events must be correlated in order to assure that both peers hold the *same* key. This can only be done by inspecting one of the certificates of message 7, so as to make A explicit. Then, Theorem 11.3.6 specifies the form of the session key that is output by B's card (Theorem 11.3.21).

Theorem 11.3.21 (SR_B_keydist_to_A). *If B is not the Spy, A and B's cards are not illegally usable and evs contains*

Outputs (Card B) B {|Nonce Nb, Key Kab, $Cert$,
              Crypt(pairK(A, B))(Nonce Nb)|}   *and*

Gets B (Crypt(pairK(A, B))(Nonce Nb))

then evs is such that

Key $Kab \in$ analz(knows A evs).

No stronger result than this can be envisaged because there exists no protocol message that binds the session key with both of its peers. Can B inspect any of the certificates of message 7? Or, can B's card inspect that of the last message? Both answers being negative, Theorem 11.3.21 turns out to be applicable by neither B nor his card, which seems a poor outcome for the protocol.

11.4 Verifying Shoup-Rubin-Bella

Omitting B's name from message 2 of the public-key Needham-Schroeder protocol led to Lowe's well-known attack [107]. Although public-key cryptography attempted to enforce confidentiality of the nonces, the Spy could intercept the messages while interleaving two sessions, learn an important nonce, and violate the authentication of the initiator to the responder. With Shoup-Rubin, the secure contexts between agents and smartcards prevent this. However, since the card data buses are not reliable (§10.1.1), when lack of explicitness affects the card outputs, the agents cannot distinguish which protocol session a single output belongs to, which seems realistic.

Abadi and Needham demonstrate that lack of explicitness may crucially affect the interpretation of a message: "The names relevant for a message can sometimes be deduced from other data and from what encryption keys have been applied. However, when this information cannot be deduced, its omission is a blunder with serious consequences" [7, §4].

As mentioned above, message 7 of Shoup-Rubin cannot inform B of A's identity both because the session key does not state its peers and because the two certificates cannot be decrypted by any agents. Nor can message 10 inform A of B's identity. Moreover, while message 7 quotes the nonce Nb that is used to build the session key, message 10 fails to do so. These components cannot be learnt from the underlying transport protocol. Therefore, upon reception of an instance of message 10, agent A cannot derive the complete form of the instance of message 7 sent during that session.

However, messages 6 and 9 quote the identity of the respective, intended peer. So, it could be argued that, should the card data buses be reliable, the calling agent could store the identity of the peer until the card returns, and associate the session key just received with that peer. Nevertheless, messages 7 and 10 violate the explicitness principles that have been mentioned throughout this chapter. Indeed, we have shown how they weaken the protocol goals when the extra assumption that the card data buses are reliable is not made.

These considerations suggest updating messages 7 and 10 to design a variant protocol, which we call Shoup-Rubin-Bella. It only differs from the original protocol in the components underlined in Figure 11.15. The formal protocol model is easy to update accordingly.

$$7. \quad C_b \to B : \quad Nb, \underline{A}, Kab, \{\!| Na, Nb |\!\}_{\pi_{ab}}, \{\!| Nb |\!\}_{\pi_{ab}}$$
$$10. \quad C_a \to A : \quad \underline{B, Nb}, Kab, \{\!| Nb |\!\}_{\pi_{ab}}$$

Fig. 11.15. Shoup-Rubin-Bella protocol: fragment

In the new model, many of the theorems discussed above obtain slightly modified assertions and, crucially, assumptions that never inspect the certificates. So, the assumptions have become verifiable by the agents, signifying that the updated protocol makes the corresponding goals available to its peers. For example, Theorem 11.3.4 now (**SRB_Outpts_which_Card_10**, omitted here) can be enforced on the event

Outputs (Card A) A $\{\!|$Agent B, Nonce Nb, Key Kab, $Cert|\!\}$

In this section, *evs* is a generic trace of the model srb for the updated protocol, which comes with the file ShoupRubinBella.thy (Figure 3.1). The assertion of Theorem 11.3.6 can be stripped of one existential, so B learns the peer for the session key (Theorem 11.3.6′). One existential still constrains the form of one of the certificates, but B's knowledge is not significantly affected.

Theorem 11.3.6′ (SRB_Outpts_B_Card_form_7). *If evs contains*

Outputs (Card B) B $\{\!|$Nonce Nb, Agent A, Key Kab, $Cert1$, $Cert2|\!\}$

then, for some Na,

$Kab = \mathsf{sesK}(Nb, \mathsf{pairK}(A, B))$ *and*
$Cert1 = \mathsf{Crypt}(\mathsf{pairK}(A, B))\{\!|$Nonce Na, Nonce $Nb|\!\}$ *and*
$Cert2 = \mathsf{Crypt}(\mathsf{pairK}(A, B))($Nonce $Nb)$.

Similarly, proving an analogous (**SRB_Outpts_A_Card_form_10**, omitted here) of Theorem 11.3.7 on the event

Outputs (Card A) A {|Agent B, Nonce Nb, Key Kab, $Cert$|}

avoids the existential quantifiers in the assertion because both B and Nb are already bound. The entire authenticity argument remains unvaried, except for an analogous of Theorem 11.3.11 (**SRB_Inputs_A_Card_9_authentic**, omitted here), which obtains the expected extra component.

The unicity results continue to hold. For example, Theorem 11.3.13 must now cope with the additional components (Theorem 11.3.13′).

Theorem 11.3.13′ (SRB_Outpts_A_Card_unique_key). *If evs contains*

Outputs (Card A) A {|Agent B, Nonce Nb, Key Kab, $Cert$|}   *and*

Outputs (Card A') A' {|Agent B', Nonce Nb', Key Kab, $Cert'$|}

then

$$A = A' \quad and \quad B = B' \quad and \quad Nb = Nb' \quad and \quad Cert = Cert'.$$

An analogous of Theorem 11.3.16 (**SRB_Confidentiality_B**, omitted here) gets the simpler main assumption

Outputs (Card B) B {|Nonce Nb, Agent A, Key Kab, $Cert1$, $Cert2$|}

which is verifiable by B because the certificates are not inspected, and so does an analogous of Theorem 11.3.17 (**SRB_Confidentiality_A**, omitted here), which rests on

Outputs (Card A) A {|Agent B, Nonce Nb, Key Kab, $Cert$|}

In consequence, the peers will be able to decide, within their minimal trust, whether the session key they obtain is confidential. Hence, that goal is available to them.

Both authentication theorems are strengthened in the sense that the corresponding goals are now available to the peers. Agent A can now be informed that B and his card were present on the network and that B's card intended to communicate with A (Theorem 11.3.18′). Agent A must only verify to receive from her card a message containing four components: an agent name, a nonce, a key and a certificate. This version purposely is compact, but the form of the certificates in the conclusion can be specified (**SRB_A_authenticates_B**, omitted here).

Theorem 11.3.18′ (SRB_A_authenticates_B_bis). *If B's card is not illegally usable and evs contains*

Outputs (Card A) A {|Agent B, Nonce Nb, Key Kab, $Cert2$|}

then, for some Cert1, evs also contains

Outputs (Card B) B {|Nonce Nb, Agent A, Key Kab, $Cert1$, $Cert2$|}.

The result can be interpreted (§11.3.6) as non-injective agreement of B and his card with A and her card on Kab. This goal is available to A and also to her card.

Theorem 11.3.19 cannot be refined straight away by not inspecting the mentioned certificate, because that would hide A's identity. It can be reformulated (Theorem 11.3.19′) by introducing an appropriate Outpts event that binds the necessary message components, and then by specifying the second certificate by an analogous of Theorem 11.3.6 proved for the updated protocol. The new theorem is applicable by B, who can check the reception from the network of a certificate previously obtained from his card. Another version specifies the form of the session key and of the second certificate (**SRB_B_authenticates_A**, omitted here).

Theorem 11.3.19′ (SRB_B_authenticates_A_bis). *If B is not the Spy, A and B's cards are not illegally usable and evs contains*

Outputs (Card B) B {|Nonce Nb, Agent A, Key Kab, $Cert1$, $Cert2$|}   *and*

Gets B ($Cert2$)

then evs also contains

Outputs (Card A) A {|Agent B, Nonce Nb, Key Kab, $Cert2$|}.

This result expresses non-injective agreement of A and her card with B on Kab. It is available to B but not to his card.

Theorem 11.3.20 can be enforced on the same assumptions as those of Theorem 11.3.18′, and so becomes applicable by A (Theorem 11.3.20′).

Theorem 11.3.20′ (SRB_A_keydist_to_B). *If B's card is not illegally usable and evs contains*

Outputs (Card A) A {|Agent B, Nonce Nb, Key Kab, $Cert2$|}

then evs is such that

Key $Kab \in$ analz(knows B evs).

Similarly, the assertion of Theorem 11.3.21 can be proved on the assumptions of Theorem 11.3.19′ and become applicable by B. The resulting theorem (**SRB_B_keydist_to_A**, omitted here) along with Theorem 11.3.20′ signifies that the Shoup-Rubin-Bella protocol makes key distribution available to both its peers.

12. Modelling Accountability

A class of recent security protocols aim at their goals even if the peers misbehave. This is a fundamental change to the threat model, and demands a novel design strategy called accountability. The Inductive Method only requires small extensions to manage accountability.

Classical security protocols establish secure communications over insecure networks. Typically they assure that no attacker can obtain sensitive information or impersonate another person. A protocol protects Alice and Bob, who trust one another, from hostile parties. This scenario is inappropriate when Alice does not even know Bob, let alone trust him. Purchasing goods over the Internet requires trusting the merchant with your credit card details, even if a protocol such as SSL protects against outsiders.

Preliminary registration is an attempt to strengthen trust. People who wish to participate must first enroll with an authority. Protocols that employ registration include SET [37] and Visa 3-D Secure [163]. Registration gives Alice some confidence in Bob — since he can present signed credentials — but it does not change the security framework. Alice still must trust Bob. We have studied this strategy extensively within the analysis of the SET registration protocols [36], which is not part of this book.

Accountability is a protocol design strategy that is meant to reduce the need for trust [169]. For example, the non-repudiation protocol of Zhou and Gollmann [171] aims at transmitting a message while ensuring that neither the sender nor the receiver can deny taking part in the process. Each of them in fact receives sufficient evidence to prove the other's participation. Another example is the certified e-mail protocol by Abadi et al. [4], which similarly assures that an e-mail is delivered if and only if its sender gets the return receipt. Both protocols are intended to achieve their goals even if the other party misbehaves: Alice need not trust Bob, and vice versa. These protocols will be analysed extensively in the next chapter. Crispo's delegation protocol [67], which is not studied in this book, is another simple instance of the accountability strategy of design [43].

This chapter presents techniques for modelling and verifying *accountability protocols*. Correctness of accountability protocols involves two concepts.

- *Validity of evidence*: an agent is given evidence sufficient to convince a third party of his peer's participation in the protocol.

– *Fairness*: both agents obtain the promised items, or neither of them does [17].

Proving the new properties required the development of novel methods for proof and especially for specification. Allowing the peer to be the adversary could make proofs excessively complicated. To keep proofs simple, the guarantees must be expressed with care. We must also formalise various forms of *secure channels*: channels that satisfy properties such as authentication, confidentiality or guaranteed delivery. Many accountability protocols rely upon secure channels, which might be implemented by running another security protocol: we thus arrive at the concept of *higher-level protocols*. Accountability is often described in terms of evidence that can be presented to a judge. We model the evidence only, not the judge, assuming that the judge holds the evidence as soon as it exists.

This chapter pragmatically introduces the challenges that the new class of protocols puts to formal analysis in general (§12.1). Then it discusses how to face the challenges using the Inductive Method (§12.2).

12.1 Challenges for Formal Analysis

The verification of security protocols appears to be mature, but we expected new challenges from accountability protocols. As a starting point, it was necessary to find suitable formalisations and proof methods for the goals of non-repudiation and certified e-mail delivery. Arguably, new security goals are increasingly complex, and hence increasingly difficult to formalise and prove. But we managed to face this very challenge rather easily (§12.1.1).

Accountability protocols may assume an available SSL channel. Coping with this assumption from the formal verification standpoint inspired unexpected considerations and raised extra concerns. Designing a security protocol presupposing that the goals of other security protocols are available follows a hierarchical design strategy that entails the notion of higher-level security protocol. We spell out this notion below (§12.1.2) but find it worthy of further investigation, independently from accountability protocols. However, our presentation respects the temporal thread of our research.

12.1.1 Formalising and Verifying the Novel Goals

If we aim at verifying new security goals, we must first solve the problem of their formalisation. The accountability goals appeared to be intrinsically different from common goals such as confidentiality and authentication, which we knew how to prove (see previous chapters). At the time, an eminent example of formal analysis of Zhou-Gollmann's protocol by pen and paper already existed [144], but our aim was to find out whether that protocol would be amenable to mechanised formal analysis. No attempt of formal analysis of the protocol by Abadi et al. was available (Abadi and Blanchet's analysis [2] was yet to be published). So, the available literature could not be of much help.

Once the goals are formalised, it is necessary to develop adequate methods for proving them. However, the phases of formalisation and proof are never separated. Inductive proof methods are appropriate for proving safety properties. So, the main task was to find a formalisation of the new goals as safety properties, whose feasibility was not at all obvious in the beginning. Once this had been achieved, it was pleasing to realise that the existing proof methods would scale up fairly easily (§12.2.1).

12.1.2 Challenges from Higher-Level Protocols

Classical security protocols, such as those from Clark and Jacob's library [61], aim at establishing various security goals. All those protocols are designed on the assumption that underlying transport protocols are available to carry out the delivery of cryptographic messages. Sometimes, a specific transport protocol, such as FTP, is assumed.

Many recent security protocols rely on some other security protocol as a primitive. For example, the fair exchange protocol by Asokan et al. [17] presupposes that the peers authenticate each other by means of an authentication protocol before they engage in a transaction whereby they commit to a contract. Likewise, the certified e-mail protocol requires the establishment of an SSL channel between the e-mail recipient and the trusted third party. As novel security goals become necessary, some of these protocols may be required by upcoming security protocols targeted at the novel goals.

In general, any kind of construction customarily makes use of existing constructions, and this process can be iterated hierarchically. Applying the technique of hierarchical design to hardware development has been common practice for decades. It appears that the same technique is starting to be applied to security protocols. For example, presupposing that the goals of SSL are available to achieve additional security goals is hierarchical design. This setting inspires our hierarchical view of protocols, specified by the following definition.

Definition 12.1.1 (Higher-Level Protocols).

- A 0th-level protocol *is a protocol that transports messages and uses no cryptography.*
- An ith-level protocol, $i \geq 1$, *is a protocol that uses cryptography, $(i\text{-}1)^{th}$-level and 0th-level protocols, and, if $i > 2$, possibly some j^{th}-level protocols for any j such that $j = 1, \ldots, i - 2$.*

A 0th-level protocol is a transport protocol, while any i^{th}-level protocol, $i \geq 1$, is a security protocol as such, namely meant to accomplish some security goals. In particular, a first-level protocol is a classical security protocol, such as Kerberos or SSL. An ith-level protocol uses some cryptographic primitives of its own, an $(i-1)^{\text{th}}$-level protocol and a transport protocol. It may also use certain protocols of levels between 1 and $i-2$ if they are defined. The word "use" is purposely left generic because it is unnecessary in this context to distinguish between interleaved and sequential composition.

Definition 12.1.1 produces the protocol hierarchy in Figure 12.1. To anticipate the next chapter's treatment, the protocol by Abadi et al. clearly is a second-level protocol, while that by Zhou-Gollmann's is a first-level protocol. The latter is indeed simpler because it presupposes no other security protocol. It would become a second-level protocol if we replaced the final FTP-get steps by transmissions over an SSL channel, for example. It would remain a first-level protocol if we replaced the FTP-get steps by transmissions over a resilient channel.

0th-level : protocols that transport messages and use no cryptography.

1st-level : protocols that use cryptography and 0th-level protocols.

2nd-level : protocols that use cryptography, 1st-level and 0th-level protocols.

3rd-level : protocols that use cryptography, 2nd-level and 0th-level protocols, and possibly also 1st-level protocols.

4th-level : protocols that use cryptography, 3rd-level and 0th-level protocols, and possibly also 1st-level or 2nd-level protocols.

5th-level : protocols that use cryptography, 4th-level and 0th-level protocols, and possibly also 1st-level or 2nd-level or 3rd-level protocols.

. . .   . . .

Fig. 12.1. Hierarchical protocol design

Many security protocols assume the existence of a PKI, a protocol whereby each agent registers his private key with a certificate authority and publishes his public key. Strictly speaking, a PKI pushes the protocols that use it one level up in the hierarchy. Because most protocol designers consider a PKI a very basic requirement, it seems fair to ignore this detail in assigning a hierarchical level to protocols.

Designing security protocols hierarchically is risky. For example, designing a second-level protocol considering that the goals of an underlying first-level protocol are available requires a strong assumption that no interaction between the two protocols can weaken the goals of either protocol. Recent research denouncing attacks to multi-protocols confirms that assumption to be strong indeed [66]. From the security standpoint, it seems preferable to flatten the hierarchy opening up all black boxes, and hence design only first-level protocols. This is simpler for few protocols, and avoids the danger of missing potential interactions between the levels, but it becomes infeasible as many protocols are combined. Also, it is clear that the hierarchical design can be engineered more simply. We remark that it is the protocol designer, not the protocol analyser, who makes a choice between flat or hierarchical design. The analyser merely ought to conform to the designer's choice and to verify whether that choice demands assumptions that can be met in practice.

Formalising the underlying protocols. Security protocols of higher levels rely on underlying protocols to achieve their own goals. Precisely, they rely on the goals achieved by the underlying protocols. Indeed, this hierarchical design strategy treats the protocols as black boxes, and considers only the security properties they make available. Hence, a formalisation should be found for the goals of the underlying protocols, which should be introduced as assumptions rather than as actual goals to prove. This may not be obvious, depending on the adopted formalism, as a security property may have a different formalisation as an assumption rather than as a goal. For example, this will be the case with authentication; we have found simple ways to introduce authentication, confidentiality and guaranteed delivery as assumptions (§12.2.2). Other properties may require dedicated treatment.

Defining and formalising a threat model. As stated above (§3.9), the standard model for first-level protocols is Dolev-Yao's. It consists of a single attacker who monitors the entire network, reuses the intercepted messages as he pleases, and yet cannot break ciphertexts. This signifies that the attacker can tamper with the transport protocol. It is not obvious how to define a threat model for higher-level protocols in general. After significant experiments, we defined the following model for second-level protocols; this may be subject to further debate.

1. *Any agent may impersonate the Dolev-Yao Spy.* This is in the spirit of Bella and Bistarelli's threat model [30]. Modern technology and the price of skills to master it have become increasingly cheaper. Hence, it is realistic that any network agent hides malicious intentions for his own sake, without interest in colluding with others or necessity to do so.
2. *The Spy can arbitrarily establish channels by means of first-level protocols and use them.* For example, the Spy must be entitled to establish an SSL-protected channel with anyone else and send arbitrary messages on it.
3. *The Spy cannot tamper with the goals of first-level protocols.* For example, if a second-level protocol rests on an authenticated channel, then the Spy cannot interpose. If that communication is confidential, then the Spy cannot overhear.

In short, the last item assumes that a secure channel is always secure. This correspondingly differs from the threat model for first-level protocols, which sees the Spy tamper with the underlying transport protocol (she controls the traffic). However, second-level protocols assume underlying first-level protocols that have been formally verified.

Once we have defined a threat model for second-level protocols, the main task remains to formalise it within the formal method of choice. We have easily managed this with our method (§12.2.3). Nevertheless, we remark that defining a suitable threat model for higher-level protocols remains in general an open problem.

12.2 Facing the Challenges

The Inductive Method had to be extended to face the three challenges raised by accountability protocols. The process required much thought to even cope with protocols of hierarchical level as low as the second, but the actual extensions turned out to be technically simple. Yet, we expect that the current strategy to formalising second-level protocols can be easily reiterated for higher-level protocols.

12.2.1 Formalising and Verifying the Novel Goals

We have found simple formalisations for the goals of non-repudiation and certified e-mail delivery. Most importantly, we have also developed rather simple methods for proving them.

The goals of non-repudiation are that at the end of a protocol session the initiator has NRR and the responder has NRO. Given a generic trace evs of the model for a generic non-repudiation protocol, say nrp, the goals can be expressed respectively as $NRR \in$ analz(knows A evs) and $NRO \in$ analz(knows B evs). If the protocol also is fair, then either both goals are achieved or neither is. In consequence, the main goal of a fair non-repudiation protocol in a very abstract version resembles a logical equivalence. Figure 12.2 presents a template formalising fair non-repudiation at a high level of abstraction.

If evs *is a generic trace of* **nrp**, NRO *binds* A *to the sending of message* m *and* NRR *binds* B *to the reception of message* m, *then*

NRO $\in$ analz(knows B evs) *if and only if* NRR $\in$ analz(knows A evs)

Fig. 12.2. Abstract formalisation template for fair non-repudiation

Also the main goal of certified e-mail delivery can be abstractly expressed as a logical equivalence. Given an e-mail m, a sender S and an intended receiver R for m, it must be the case that R receives m if and only if S obtains the receipt RR that R received m. Figure 12.3 presents a template formalising certified e-mail delivery at a high level of abstraction, cedp being the model for a generic protocol for certified e-mail delivery.

> If **evs** *is a generic trace of* **cedp** *and* **RR** *is the return receipt for e-mail m, then*
>
> **m** ∈ **analz(knows R evs)** *if and only if* **RR** ∈ **analz(knows S evs)**

Fig. 12.3. Abstract formalisation template for certified e-mail delivery

The specific events expressing in the preconditions the roles of initiator (sender) or responder (receiver) played by the agents are abstracted away from the abstract templates. Clearly, their forms depend on the specific protocol. Most importantly, the formalisations require no agent's minimal trust to include that the agent's peer is not the Spy. It means that everyone is protected from anyone else who may try to subvert the protocol. The agent for whom the guarantee is being designed may be required to be different from the Spy; hence, she gets the guarantee at the price of acting legally himself.

Each implication of either logical equivalence is intended for a protocol participant. For example, consider Figure 12.3. The left-to-right implication reads as follows: if the receiver can derive the e-mail from the portion he sees of the network traffic on *evs*, then the sender can derive the corresponding return receipt from the portion he sees of the network traffic on *evs*. This guarantee confirms that the sender never has a disadvantage over the receiver, even if the receiver is able to derive the e-mail off-line from the analysed components. Likewise, the right-to-left implication reads as follows: if the sender can derive a return receipt, then the receiver can derive the corresponding e-mail. When the sender submits his return receipt to a judge, this guarantee counts as evidence against the receiver.

The certificates *NRO*, *NRR* and *RR* involve a digital signature or other mechanism to prevent them from being forged. Such mechanisms are invisible at the current level of abstraction. Should extra detail be required, an appeal to the synth operator, which expresses message creation, would be of help. Facts of the form *NRO* ∈ synth(analz(knows *S evs*)) could be used instead.

Given specific protocols, the abstract formalisations of the goals must be refined. Despite their nice symmetry, the conclusions are weak because the operator analz represents a potentially unlimited series of decryptions by available keys. A stronger formalisation would replace the use of analz by a reference to a specific protocol message that delivers the required item, which an honest agent expects to be given directly. Let us consider the goal in Figure 12.2, for example. We will see that the refined equivalence for Zhou-Gollmann's protocol adds knowledge of an extra message component, called *con_K*, to both sides. Likewise, let us consider the goal in Figure 12.3. The refined left-to-right implication for the protocol by Abadi et al. will say that if the receiver can compute the e-mail by an unlimited amount of work, then the sender has been given the corresponding return receipt. The refined right-to-left implication will say that if the sender can compute a return receipt by an unlimited amount of work, then the receiver has been given the corresponding e-mail. Observe that we cannot express this refined formalisation properly unless we know the precise format of the protocol messages.

Sometimes, we can prove stronger facts that no longer involve analz. For example, here is an improved version of the right-to-left implication of certified e-mail delivery.

> If a return receipt has been created at all, then the receiver has been given the corresponding e-mail.

This version does not refer to the sender. We have strengthened the guarantee while eliminating the need to reason about the sender's knowledge.

The same technique can be applied to non-repudiation, but not to the left-to-right implication of certified e-mail delivery because obviously the e-mail m will have been created. Instead, we are forced to divide that implication into two separate properties. One concerns the Spy and is proved by reasoning about analz(knows Spy evs), which we know how to do. The other concerns an honest agent, and states that if R is given the message (in the normal way) then S has been given the receipt.

Accountability protocols typically use a key to protect their transmitted message. Later, the responder (receiver) is given this key, so that she can decrypt the message. Therefore, a statement such as the above, of the form "agent has the message" can be replaced by "agent has the key to the message." We can also formalise a further guarantee, namely that the key never reaches the Spy. This guarantee, which is of obvious value to both parties, is formalised and proved much as it would be for a typical first-level protocol.

12.2.2 Formalising the Underlying Protocols

Formalising second-level protocols requires a formalisation of the goals of the underlying first-level protocols as black boxes (§12.1.2). Therefore, the main goals that we need to model are authentication and confidentiality. Other minor goals, such as guaranteed delivery, may also be of interest. Below, we describe how to model these properties abstractly in our inductive framework: in other words, how to model channels secured by first-level protocols.

Authentication. The sender identity A of a Says $A B X$ event cannot be altered in the model once that event occurs in a trace. When we formalise a first-level protocol, we must not allow an agent to inspect the sender identity; the originator of a message remains unknown unless this is conveyed by the message itself. As seen above (Chapter 8), we can formalise message reception using the Gets event, which does not mention the sender. The original formulation [133] used the event Says $A' B X$, taking care to assure that the value of A' (the true sender) was never used.

On the basis of these observations, one strategy to formalising authentication as an assumption is to allow specific Says events among the rule preconditions, and explicit reference to the sender variable among the post-conditions. For example, Says $A\,B\,X$ would signify that B can authenticate X as coming from A. This is the right way to model an authenticated channel that does not offer confidentiality, because the Spy can read X from the event Says $A\,B\,X$.

Confidentiality. What if the channel must be confidential? We could extend our definitional framework, introducing an additional event ConfSays $A\,B\,X$ for sending a message confidentially. This would require extending the definition of the function knows, which formalises each agent's knowledge. The new event ConfSays $A\,B\,X$ would make X available only to the designated recipient, B. If we include the event ConfSays $A\,B\,X$ as the precondition of a rule and allow other references to A, who is the true sender, then we have modelled an authenticated, confidential channel. If we forbid other references to A, then our confidential channel is unauthenticated. We performed some experiments using the new event but abandoned them when we realised that the original definitional framework was already sufficient to model secure channels.

The Notes event formalises an agent's changing his internal state. It has the form Notes $A\,X$, where X is a message (perhaps the result of a computation) being stored for future reference. We can formalise a confidential transmission of a message X from A to B by the specific event

$$\text{Notes}\,A\,\{\!|A, B, X|\!\} \tag{12.1}$$

Observe that the identities of the peers are stored with the actual message by convention, as is demonstrated by rule *DSLP1* in Figure 12.4. The figure shows a demo second-level protocol whose first message is protected by the security properties of a preestablished first-level protocol, for example, SSL. Its inductive model is `dslp`. Event 12.1 must be included as a precondition of a new inductive rule formalising reception of the confidential message, because reception is in general not guaranteed even on a confidential channel. The new reception rule must introduce the event

$$\text{Notes}\,B\,\{\!|A, B, X|\!\} \tag{12.2}$$

signifying that B receives X confidentially, as demonstrated by rule *Reception1st* in Figure 12.4. Event 12.2 must be included as a precondition of the rule formalising B's actions upon reception of the confidential message X. The message is therefore authenticated to arrive from the agent whose identity appears as first component of the noted message. This is demonstrated by rule *DSLP2* in Figure 12.4.

No further construct is necessary to model second-level protocols, except for the additional rule *Fake1st*, which is explained in the next section. In particular, the *Fake* rule remains unvaried. Observe that, because of our specific use of Notes to formalising the security properties, it is important to make sure that no other use of that event involves messages beginning with two agent names.

$$
\begin{array}{llll}
1. & A \xrightarrow{\text{SSL}} B & : & A, Na \\
2. & B \longrightarrow A & : & \{Na\}_{sK_B^-}
\end{array}
$$

Nil:
`[] ∈ dslp`

Fake:
`⟦ evsF ∈ dslp; X ∈ synth(analz(knows Spy evsF)) ⟧`
`⟹ Says Spy B X # evsF ∈ dslp`

Fake1st:
`⟦ evsF1 ∈ dslp; X ∈ synth(analz(knows Spy evsF1)) ⟧`
`⟹ Notes B {Agent Spy, Agent B, X} # evsF1 ∈ dslp`

Reception:
`⟦ evsR ∈ dslp; Says A B X ∈ set evsR ⟧`
`⟹ Gets B X # evsR ∈ dslp`

Reception1st:
`⟦ evsR1 ∈ dslp; Notes A {Agent A, Agent B, X} ∈ set evsR1⟧`
`⟹ Notes B {Agent A, Agent B, X} # evsR1 ∈ dslp`

DSLP1:
`⟦ evs1 ∈dslp; Nonce Na ∉ used evs1 ⟧`
`⟹ Notes A {Agent A, Agent B, Nonce Na} # evs1 ∈ dslp`

DSLP2:
`⟦ evs2 ∈ dslp;`
`  Notes B {Agent A, Agent B, Nonce Na} ∈ set evs2 ⟧`
`⟹ Says B A (Crypt (priSK B) (Nonce Na)) # evs2 ∈ dslp`

Fig. 12.4. Example second-level protocol and corresponding inductive model

Guaranteed delivery. Other minor goals of first-level protocols can be formalised using similar techniques. For example, distribution of a session key to a pair of agents can be formalised by an inductive rule that gives both agents Notes events containing a key, with a precondition that the key is fresh. Another goal is guaranteed delivery, which can be implemented by sending a message repeatedly until there is an acknowledgement. This goal can be easily formalised by introducing the event for receiving a message at the same time as the event for sending it. It is even simpler just to introduce the former event: the sender magically causes the message to reach the recipient. This formalisation will be extensively used below (§13.2.1) to model the certified e-mail protocol by Abadi et al. Observe that with either approach to guaranteed delivery, no reception rule in the style of *Reception1st* (Figure 12.4) is necessary.

Notes events are affected by a detail of our model, the set bad of *compromised agents*. These are honest agents that have somehow come under the control of the Spy, perhaps through a security lapse. The Spy knows their private keys and can read their Notes. This detail is consistent with our use of Notes above, since the Spy can be expected to grab anything that a compromised agent receives, even via a secure channel. The model does not constrain bad other than asserting that the Spy belongs to it.

It must be remarked that traditional first-level protocols are designed to protect honest agents from the rest of the world, and a typical guarantee will hold provided that both peers are uncompromised. However, an agent A who executes a non-repudiation protocol or a certified e-mail protocol with B requires the protocol goals *especially* if B is bad. In particular, B might even be the Spy, who can send arbitrary messages built from known components.

Let us emphasise what we have provided here, namely a formalisation of the security properties assumed by a second-level protocol. The specific underlying first-level protocol that achieves those properties has not been considered. Following a hierarchical verification strategy, it is irrelevant what is hidden inside the black box (whether SSL or Kerberos, for example) that is assumed to provide the goals. The same approach can be taken to formalising the security properties assumed by protocols of hierarchical level higher than the second.

12.2.3 Formalising a Threat Model

We have seen that the threat model formalised in the Inductive Method is a Dolev-Yao Spy (§3.9). In principle, it is not obvious what the threat model for second-level protocols is. We have defined what we believe is a realistic one (§12.1.2), briefly recalled here.

1. Any agent may impersonate the Dolev-Yao Spy.
2. The Spy can arbitrarily establish channels by means of first-level protocols and use them.
3. The Spy cannot tamper with the goals of first-level protocols.

Condition 1 is easy to implement in the Inductive Method. It may be seen in Figure 12.4 that no rule is conditioned to agents other than the Spy. So, its implementation does not influence the formalisation of a protocol. However, it affects the formalisation of the guarantees, which must now be proved admitting that the Spy can hide behind any of the involved peers. This care, which we do take below, complicates the proofs with respect to their earlier versions where we did not take it [35].

Conditions 2 and 3 together mean that the Spy can send messages at will on channels she establishes using first-level protocols, but cannot interfere with any such channels established by other agents. Modelling first-level protocols using Notes events, as explained above, yields this threat model for free — namely with no changes to our definitional framework — except for the following detail.

The extra rule *Fake1st* is needed to allow the Spy to send arbitrary messages on channels established by first-level protocols (Figure 12.4). Indeed, the rule introduces an event of the form

$$\text{Notes } B \, \{\!| \text{Spy}, B, X |\!\}$$

which formalises B's reception of the Spy's arbitrary message X on an authenticated channel. This adds realism to the model. If an agent authenticates himself to another, it does not mean that he will act honestly: he can still attempt cheating by sending arbitrary messages to the other agent.

13. Verifying Two Accountability Protocols

A comparative verification of two important protocols designed using the accountability strategy is presented. One protocol, due to Zhou and Gollmann, is for non-repudiation and the other one, due to Abadi et al., is for certified e-mail delivery.

An *accountability protocol* gives an agent lasting evidence, typically digitally signed, about actions performed by his peer. Many authentication mechanisms fail to meet this requirement: the reply to an encrypted nonce challenge proves an agent's presence to the recipient but to nobody else. The protocol should meet its objectives to an honest agent even if the peer misbehaves. This chapter presents a comparative analysis of a non-repudiation protocol [171] and a certified e-mail protocol [4]. Both protocols have two peers and a trusted third party TTP.

The results presented here supersede our previous work, especially for the non-repudiation protocol [42], which was the first to be studied, though in a limited model. It was only later that we developed the new specification and verification methods described in the previous chapter. We first adopted them to analyse the certified e-mail protocol [35], and then to reexamine the non-repudiation protocol. For both protocols, it is now proved that an agent who obeys the rules is also protected from a peer who is possibly cheating. A more compact version of this conjunct analysis is also published [44].

We have found both protocols to be correct: they are fair and they deliver valid evidence. More precisely, the non-repudiation protocol is fair in the sense that the initiator gets non-repudiation of receipt if and only if the responder gets non-repudiation of origin. Both pieces of evidence are proved valid. Along with his evidence, the responder also gets the message that the initiator intended to send him. In comparison, the certified e-mail protocol achieves slightly weaker goals. It is fair in the sense that the initiator gets non-repudiation of receipt (a "return receipt") if and only if the responder gets the e-mail. That receipt is proved valid. However, the responder gets no evidence for non-repudiation of origin. The e-mail itself does not suffice for this purpose.

The typical accountability setting sees the sender's intention to transmit a message m to B. She encrypts it using a symmetric key, K; the ciphertext $c = m_K$ is her *commitment* to the session with the receiver. Recall that the private signature key of an agent X is indicated as sK_X^-, while the signature of a message y by key sK_X^- is $\{\!|y|\!\}_{sK_X^-}$. The public encryption key of an agent X is indicated as eK_X, while the encryption of a message y by key eK_X is $\{\!|y|\!\}_{eK_X}$. Our notation (§3.6) makes no distinction between the operations of symmetric encryption, asymmetric encryption and signature, because the type of key suffices to discern.

The structure of this chapter is simple. First, we present the non-repudiation protocol, its model and its verified properties (§13.1). Then, we do the same with the certified e-mail protocol (§13.2). Finally, we compare and contrast them (§13.3).

13.1 The Non-repudiation Protocol

The non-repudiation protocol (Figure 13.1) uses a lightweight TTP whose effort is independent of the size of the transmitted message. A unique label, L, identifies the session between A and B. It concerns two types of evidence. Non-repudiation of origin (NRO) proves the participation of the initiator A, while non-repudiation of receipt (NRR) proves the participation of the responder B. Flags such as f_{nro} express the non-repudiation meaning of a certificate.

Abbreviations

$$c = m_K$$
$$NRO = \{\!|f_{nro}, B, L, c|\!\}_{sK_A^-}$$
$$NRR = \{\!|f_{nrr}, A, L, c|\!\}_{sK_B^-}$$
$$sub\_K = \{\!|f_{sub}, B, L, K|\!\}_{sK_A^-}$$
$$con\_K = \{\!|f_{con}, A, B, L, K|\!\}_{sK_{TTP}^-}$$

Steps

1.	A	$\longrightarrow$	B	: f_{nro}, B, L, c, NRO
2.	B	$\longrightarrow$	A	: f_{nrr}, A, L, NRR
3.	A	$\longrightarrow$	TTP	: f_{sub}, B, L, K, sub_K
4.	B	$\xleftarrow{\text{FTP}}$	TTP	: $f_{con}, A, B, L, K, con\_K$
5.	A	$\xleftarrow{\text{FTP}}$	TTP	: $f_{con}, A, B, L, K, con\_K$

Fig. 13.1. Non-repudiation protocol by Zhou and Gollmann

The protocol is easy to describe. First (step 1), A picks a symmetric key K and a label L, and encrypts m with K to form c. Then, A signs f_{nro}, B, L, c to yield NRO, which she sends to B. In response (step 2), B verifies A's signature, signs f_{nrr}, A, L, c and sends the resulting NRR to A. Then (step 3), A lodges K with TTP by sending sub_K, which is f_{sub}, B, L, K signed with her private signature key.

If TTP can verify A's signature, it signs f_{con}, A, B, L, K producing con_K, which it makes available in its public directory. This step binds the key K to the session between A and B labelled L. Finally (steps 4 and 5), A and B download con_K from TTP using the File Transfer Protocol (FTP); the protocol assumes that this download will eventually succeed.

The authors state that "In practice, we will not want TTP to store message keys forever. We could set a deadline T to limit the time con_K and K can be accessed by the public" [171, §5.3]. This statement only implicitly addresses the question of an off-line audit trail, which might be considered obligatory. Without a full audit trail, the protocol suffers a form of replay attack: the initiator can reuse the supposedly unique session label, using evidence from a past run to "prove" the responder's participation in a recent session [87]. This attack seems unrealistic, as auditing is notoriously fundamental to security. Gollmann has rejected it in a private communication. as is based on traces, our model assumes that a full audit trail is always available, and hence will not signal irregularities of this form.

The protocol aims at providing each party with evidence to prove the other's participation. The evidence for A consists of NRR and con_K, while that for B consists of NRO and con_K. Making con_K part of the evidence is particularly important: it assures fairness, since TTP releases this item to both parties simultaneously.

Let us informally analyse how to resolve disputes. If A holds con_K and NRR, then she has completed a run with B, who has accepted the commitment c and should be able to download the decryption key from TTP. Similarly, if B holds con_K and NRO, then A cannot deny having sent c as a commitment bound to label L. Of course, such arguments are unconvincing as they stand; we need formal verification.

13.1.1 Model

The protocol model is the set of traces zg, whose inductive definition is represented in Figure 13.2. It can be found in file ZhouGollmann.thy (Figure 3.1). Rules Nil and Fake are standard. Rules ZG1, ZG2, ZG3 and ZG4 respectively model the legitimate protocol steps. In particular, to initiate the protocol with B, agent A chooses a fresh label in rule ZG1. "Labels have to be unique to create the link between commitment and key" [171, §5.2], so we decide to model them as random numbers, namely as nonces. Therefore, our labels are independent of the messages — we do not study more detailed computations of labels. Because A sends the message m in an encrypted form, she must choose a cryptographic key. Rule *ZG1* leaves her free to choose any key, even an old one; we merely assume that she cannot pick asymmetric keys.

We highlight the important certificates by defining them in the premises using equations, and using the names so defined in the conclusions. When a certificate is defined in the premises of a rule, the rule only applies for a certificate of the specified form; informally, the agent verifies it. For example, B must check that *NRO* in rule *ZG2* is signed by A in order to confirm that she is the sender of the message just received. Likewise, A must check that *NRR* in rule *ZG3* is signed by B.

Rule *ZG4* models TTP's preparation of the key confirmation *con_K*. TTP verifies the signature on *sub_K* to confirm the identities of the other agents. All the components needed to verify the signature are available. The installation of *con_K* in TTP's public directory is modelled by a Notes TTP event. Also the Spy must be able to download it, so two events are needed. Because TTP is uncompromised, the Notes TTP event keeps *con_K* from the Spy, while a Says TTP Spy event explicitly gives it to her. This rule terminates the protocol model: it is unnecessary to formalise the peers' retrieval of *con_K*.

13.1.2 Verification

For verifying this protocol, an additional definition is necessary: the set of *broken* agents. A compromised agent's digital signatures are worthless, but the Spy's own signatures must be considered valid. The set broken therefore includes all compromised agents other than the Spy.

$$\text{broken} \triangleq \text{bad} \setminus \{\text{Spy}\}$$

If an agent is broken, then the Spy has his keys and can impersonate him freely, but the agent certainly is not the Spy. Conversely, if an agent is unbroken, then the Spy cannot impersonate him, but he can be the Spy himself. He can misbehave sending arbitrary messages as is appropriate for accountability protocols. Hence, with accountability protocols, an agent can assume his peer to be unbroken as part of his minimal trust. This is an important element of the threat model advanced in the previous chapter. Classical goals such as confidentiality are less relevant in this context. For example, it is emblematic that in message 3 agent A actually broadcasts the key K.

In this section, *evs* is a generic trace of the formal protocol model zg.

Proving validity of evidence. Let us start with evidence *con_K* and *sub_K*. The relevant fragments of the proof script can be found in Appendix D.2. If *con_K* exists at all (as formalised by the function used), then TTP has stored it on the FTP site, where it is available to A and B (Lemma 13.1.1). This is as expressive as can be because our current model purposely does not include the actual FTP-get operations. Either agent, possessing *con_K*, can use this guarantee to show that the peer has access to

Nil:
[] ∈ zg

Fake:
⟦ evsF ∈ zg; X ∈ synth (analz (knows Spy evsF)) ⟧
⟹ Says Spy B X # evsF ∈ zg

Reception:
⟦ evsR ∈ zg; Says A B X ∈ set evsR ⟧ ⟹ Gets B X # evsR ∈ zg

ZG1:
⟦ evs1 ∈ zg; Nonce L ∉ used evs1; c = Crypt K (Number m);
  K ∈ symKeys;
  NRO = Crypt (priK A) {|Number f_nro, Agent B, Nonce L, c|} ⟧
⟹ Says A B {|Number f_nro, Agent B, Nonce L, c, NRO|} # evs1 ∈ zg

ZG2:
⟦ evs2 ∈ zg;
  Gets B {|Number f_nro, Agent B, Nonce L, c, NRO|} ∈ set evs2;
  NRO = Crypt (priK A) {|Number f_nro, Agent B, Nonce L, c|};
  NRR = Crypt (priK B) {|Number f_nrr, Agent A, Nonce L, c|} ⟧
⟹ Says B A {|Number f_nrr, Agent A, Nonce L, NRR|} # evs2 ∈ zg

ZG3:
⟦ evs3 ∈ zg; c = Crypt K m; K ∈ symKeys;
  Says A B {|Number f_nro, Agent B, Nonce L, c, NRO|} ∈ set evs3;
  Gets A {|Number f_nrr, Agent A, Nonce L, NRR|} ∈ set evs3;
  NRR = Crypt (priK B) {|Number f_nrr, Agent A, Nonce L, c|};
  sub_K = Crypt (priK A) {|Number f_sub, Agent B, Nonce L, Key K|} ⟧
⟹ Says A TTP {|Number f_sub, Agent B, Nonce L, Key K, sub_K|}
       # evs3 ∈ zg

ZG4:
⟦ evs4 ∈ zg; K ∈ symKeys;
  Gets TTP {|Number f_sub, Agent B, Nonce L, Key K, sub_K|} ∈ set evs4;
  sub_K = Crypt (priK A) {|Number f_sub, Agent B, Nonce L, Key K|};
  con_K = Crypt (priK TTP) {|Number f_con, Agent A, Agent B,
                             Nonce L, Key K|} ⟧
⟹ Says TTP Spy con_K
   # Notes TTP {|Number f_con, Agent A, Agent B, Nonce L, Key K, con_K|}
       # evs4 ∈ zg

Fig. 13.2. Inductive model of the non-repudiation protocol

con_K, and therefore to the key K. Since con_K is equally available to both parties, this lemma also expresses an aspect of fairness. Technically speaking, it has the form of an authenticity guarantee, so the proving experience accumulated thus far is useful.

Theorem 13.1.1 (ZG_con_K_validity). *If con_K exists on evs and*

$con\_K =$
Crypt(priK TTP){|Number f_{con}, Agent A, Agent B, Nonce L, Key K|}

then evs contains

Notes TTP {|Number f_{con}, Agent A, Agent B, Nonce L, Key K, con_K|}.

The proof is a simple induction: since con_K is signed by TTP, who is uncompromised, a regularity lemma tells Isabelle's simplifier that it is an integral certificate. Hence, rule **ZG4** must have been executed. The proof script is six lines long. The first three lines, which are routine, set up the induction. The fourth line is also routine, applying theorem **ZG_ZG2_msg_in_parts_spies** (omitted here), causing Isabelle to note that the encrypted message is available to the Spy as it was sent in the network. Then, it simplifies all subgoals arising from the induction. Only two subgoals survive the simplification. They are proved routinely by **blast** with the help of technical lemmas available in **Message.thy** (Figure 3.1) concerning the relation between fake messages and the parts primitive.

An analogous theorem can be proved about sub_K. If TTP receives it within an instance of message 3, then it was sent by A (Theorem 13.1.2). Its conclusion holds even if A is the Spy, which is allowed by the assumption that A is unbroken.

Theorem 13.1.2 (ZG_sub_K_validity). *If A in unbroken and*

$sub\_K =$ Crypt(priK A){|Number f_{sub}, Agent B, Nonce L, Key K|}

and evs contains

Gets TTP {|Number f_{sub}, Agent B, Nonce L, Key K, sub_K|}

then evs contains

Says A TTP {|Number f_{sub}, Agent B, Nonce L, Key K, sub_K|}.

The proof method is new. Thanks to the reception invariant, someone sent that instance of message 3 to TTP. Then, the assumption that A is unbroken produces two subgoals: one has A uncompromised, the other has her the Spy (Figure 13.3).

The first subgoal can be solved by a conventional authenticity theorem (**ZG_sub_K_validity_good**, omitted here) stating that because sub_K is in the traffic, and because it appears to have originated with an uncompromised A, it did. The second can be solved by a less conventional, weaker result that

```
1. ⟦ evs ∈ zg;
     Says Aa TTP
       {|Number f_sub, Agent B, Nonce L, Key K,
          Crypt (priK A) {|Number f_sub, Agent B, Nonce L, Key K|}|}
       ∈ set evs;
     A ∉ bad ⟧
   ⟹ Says A TTP
        {|Number f_sub, Agent B, Nonce L, Key K,
          Crypt (priK A) {|Number f_sub, Agent B, Nonce L, Key K|}|}
        ∈ set evs;

2. ⟦ evs ∈ zg;
     Says Aa TTP
       {|Number f_sub, Agent B, Nonce L, Key K,
          Crypt (priK Spy) {|Number f_sub, Agent B, Nonce L, Key K|}|}
       ∈ set evs ⟧
   ⟹ Says Spy TTP
        {|Number f_sub, Agent B, Nonce L, Key K,
          Crypt (priK Spy) {|Number f_sub, Agent B, Nonce L, Key K|}|}
        ∈ set evs;
```

Fig. 13.3. Proving validity of $sub\,K$ for the non-repudiation protocol

makes no assumption on A. It states that sub_K originated either with A or with the Spy (Lemma 13.1.1). The proof is trivial induction since only the Spy would use another agent's key. It can be seen that the form of all message components but the last one does not need to be specified.

Lemma 13.1.1 (ZG_sub_K_sender). *If evs contains*

Says A' TTP $\{|f, b, l, k, \mathsf{Crypt}(\mathsf{priK}\,A)\,X|\}$

then A' is either A or the Spy.

The same method can be used to prove validity of the main pieces of evidence. The relevant fragments of the proof script can be found in Appendix D.1. In particular, NRO is valid (Theorem 13.1.3).

Theorem 13.1.3 (ZG_NRO_validity). *If A is unbroken and*

$NRO = \mathsf{Crypt}(\mathsf{priK}\,A)\{|\mathsf{Number}\,f_{nro}, \mathsf{Agent}\,B, \mathsf{Nonce}\,L, c|\}$

and evs contains

Gets B $\{|\mathsf{Number}\,f_{nro}, \mathsf{Agent}\,B, \mathsf{Nonce}\,L, c, NRO|\}$

then evs contains

Says $A\,B$ $\{|\mathsf{Number}\,f_{nro}, \mathsf{Agent}\,B, \mathsf{Nonce}\,L, c, NRO|\}$.

The three theorems seen so far confirm validity of B's evidence. If he exhibits con_K, then by Theorem 13.1.1 it originated with TTP. This implies that TTP received sub_K inside a valid instance of message 3

(**ZG_Notes_TTP_imp_Gets**, omitted here), because TTP only works reliably. Finally, by Theorem 13.1.2, we have that it was A who submitted the key K bound to the label L. If B also exhibits NRO, then by Theorem 13.1.3, he can assert that A submitted the commitment c bound to the label L. The label binds the commitment to the key; hence, the theorems together confirm A's intention to send the plaintext message contained in c.

Turning to A's evidence, we can prove that NRR is valid (Theorem 13.1.4) in a similar fashion. Precisely, any instance of NRR that appears to come from B actually did. As usual, the assumption that B is unbroken allows him to be the Spy.

Theorem 13.1.4 (ZG_NRR_validity). *If B is unbroken and*

$$NRR = \mathsf{Crypt}(\mathsf{priK}\,B)\{\!|\mathsf{Number}\,f_{nrr}, \mathsf{Agent}\,A, \mathsf{Nonce}\,L, c|\!\}$$

and evs contains

$$\mathsf{Gets}\,A\,\{\!|\mathsf{Number}\,f_{nrr}, \mathsf{Agent}\,A, \mathsf{Nonce}\,L, NRR|\!\}$$

then evs contains

$$\mathsf{Says}\,B\,A\,\{\!|\mathsf{Number}\,f_{nrr}, \mathsf{Agent}\,A, \mathsf{Nonce}\,L, NRR|\!\}.$$

We now have all guarantees confirming validity of A's evidence: *con_K* and NRR. When A exhibits NRR, she can assert by Theorem 13.1.4 that B holds the ciphertext c. When A exhibits *con_K*, she shows by Theorem 13.1.1 that it was available for B to download from TTP. Therefore, B can have access to m.

Proving fairness. The fairness guarantees protect an agent who follows the protocol from one who does not. The agent receiving the guarantee must be uncompromised, but no assumption is made about the peer. Since Theorem 13.1.1 already states that *con_K* is equally available to both parties, we only have to prove fairness for NRO and NRR. For the sake of readability we prefer to word the fact "$X \in$ used *evs*" as "X exists on *evs*" in this presentation, while the relevant fragments of the proof script can be found in Appendix D.3.

Here is a guarantee of fairness for B: if NRR exists at all, then B, who must be uncompromised, holds NRO (Theorem 13.1.5). The proof is a straightforward induction.

Theorem 13.1.5 (ZG_B_fairness_NRR). *If B is uncompromised, NRR exists on evs and*

$$NRR = \mathsf{Crypt}(\mathsf{priK}\,B)\{\!|\mathsf{Number}\,f_{nrr}, \mathsf{Agent}\,A, \mathsf{Nonce}\,L, c|\!\}$$

and

$$NRO = \mathsf{Crypt}(\mathsf{priK}\,A)\{\!|\mathsf{Number}\,f_{nro}, \mathsf{Agent}\,B, \mathsf{Nonce}\,L, c|\!\}$$

then evs contains

Gets B {Number f_{nro}, Agent B, Nonce L, c, NRO}.

Fairness for A has a slightly different form: if con_K and NRO exist, then A holds NRR (Theorem 13.1.6). It can be seen how con_K gives fairness to A, who otherwise would be at a disadvantage because the first message gives evidence to B.

Theorem 13.1.6 (ZG_A_fairness_NRO). *If A is uncompromised, con_K exists on evs and*

$con\_K =$
Crypt(priK TTP){Number f_{con}, Agent A, Agent B, Nonce L, Key K}

and

$NRR = $ Crypt(priK B){Number f_{nrr}, Agent A, Nonce L, c}

and

$NRO = $ Crypt(priK A){Number f_{nro}, Agent B, Nonce L, c}

and evs is such that

$NRO \in$ parts(knows Spy evs)

then evs contains

Gets A {Number f_{nrr}, Agent A, Nonce L, NRR}.

The proof is much more complicated than that of the corresponding property for B. It uses a lemma stating that A only sends message 3 after she has received NRR (**ZG_sub_K_imp_NRR**, omitted here). She recognises the correct NRR by the label L, which she is required to choose uniquely to identify the transaction. This uniqueness is important. If A attempts to cheat by reusing transaction identifiers, as suggested by Gürgens and Rudolph [87], she runs the risk of accepting the wrong transaction.

An agent can assume his peer to be unbroken as part of his minimal trust. It is perfectly realistic for accountability protocols, as unbroken agents can be the Spy. The assumptions of the guarantees presented so far can be verified by their intended beneficiaries. In consequence, the non-repudiation protocol makes its goals of validity of evidence and fairness available to its peers.

13.2 The Certified E-mail Protocol

Abadi et al. [4] have designed a realistic protocol for certified e-mail delivery. No public-key infrastructure is necessary: TTP has signature and encryption keys, but other agents merely share a password with TTP. Agent R's password is indicated pwd_R. In common with the previous protocol, TTP is lightweight and its effort is independent of the e-mail size; moreover, this

TTP is stateless. A challenge-response mechanism authenticates the receiver to the sender, who must agree beforehand on some acknowledgement function linking a challenge q to its response r.

As in the non-repudiation protocol, the sender forms a commitment by encrypting his e-mail with a symmetric key, K. He attaches K, encrypted with TTP's public encryption key. The recipient forwards the message to TTP in order to obtain K. (In the non-repudiation protocol, the sender lodges the key directly with TTP.) TTP releases the key and simultaneously releases a certificate documenting the transaction to the sender.

Abbreviations

$$c = m_K$$
$$h_S = \mathsf{Hash}(cleartext, q, r, c)$$
$$h_R = \mathsf{Hash}(cleartext, q, r, c)$$
$$S2TTP = \{\!|S, BothAuth, K, R, h_S|\!\}_{eK_{\mathsf{TTP}}}$$
$$RR = S2TTP_{sK_{\mathsf{TTP}}^-}$$

Steps

1.	S	$\longrightarrow$	R	: $\mathsf{TTP}, c, BothAuth, cleartext, q, S2TTP$
2.	R	$\xrightarrow{\text{SSL}}$	TTP	: $S2TTP, pwd_R, h_R$
3.	TTP	$\xrightarrow{\text{SSL}}$	R	: K, h_R
4.	TTP	$\longrightarrow$	S	: RR

Fig. 13.4. Certified e-mail protocol by Abadi et al.

We present the full version of the protocol (Figure 13.4), where the receiver authenticates to both the sender and TTP. This is the strongest authentication option, signaled by tag *BothAuth*. "The part *cleartext* is a header that asks R to read the certified e-mail, perhaps explaining what is in the message"[4, §3.2]. In step 1, the sender S sends the receiver R the encrypted e-mail c, a challenge q, and a certificate for TTP, called $S2TTP$. The certificate is encrypted under TTP's public encryption key and contains two important components: the symmetric key K that protects c, and a hash linking c to the required response, r. Recall that R and S must have already agreed out of band on a query-response mechanism.

In step 2, R computes the response r to the query q and includes the received ciphertext c to build the hash h_R, which he sends along with the received certificate and his password (pwd_R) to TTP on a secure channel. The authors state that security here means confidentiality and authentication, and that "in practice, such a channel might be an SSL connection" [4]. They also require guaranteed delivery, which can be implemented by sending a message repeatedly until it is acknowledged [70].

In step 3, TTP decrypts and verifies the received certificate. Then, TTP authenticates R by the password and — to check that S and R agree on the authentication mechanism — verifies that h_S found inside the ticket matches h_R. If satisfied, TTP replies to R, delivering the key found inside the ticket. This reply goes along the secure channel created in step 2.

In step 4, TTP sends a signed return receipt RR to S. Observe that RR is essentially non-repudiation of receipt (NRR). TTP must take this step jointly with the previous one, so as to be fair to both sender and receiver. If the certificate received inside the return receipt matches S's stored certificate, then S authenticates R.

In both protocols, TTP sees the symmetric key K, but not the plaintext message m. This reduces the trust in TTP, which cannot disclose the e-mails even if compromised. However, a misbehaving TTP could eavesdrop on the initial message from S to R, taking the ciphertext m_K, which he could decrypt once he knows K.

The protocol's use of encryption should prevent spies from learning m. Most importantly, the protocol "should allow a sender, S, to send an e-mail message to a receiver, R, so that R reads the message if and only if S receives the corresponding return receipt" [4, §2]. This objective is similar to that of Zhou and Gollmann, but weaker. The responder does not receive non-repudiation of origin (NRO), namely evidence that the initiator intended to send him the message. Nenadič et al. [127] have recently published an e-mail protocol that provides non-repudiation of both origin and receipt.

13.2.1 Model

The protocol model is the set of traces aghp (the authors' initials), whose inductive definition is in Figure 13.5. It can be found in file CertifiedEmail.thy (Figure 3.1). It is built according to the template given in the previous chapter in Figure 12.4. We do not model weaker authentication options than BothAuth. For authentication, R must be able to respond to a query q from S. The two agents should have agreed off-line on a series of challenge-response pairs. We choose the following implementation of responses, which allows the Spy to generate the response if R is compromised — though not if S is compromised. It is a function returning a message

response : [agent, agent, nat] $\longrightarrow$ msg

defined as

response S R q $\triangleq$ Hash{|Agent S, Key (shrK R), Nonce q|}

Message transmission over a secure channel, which is authenticated, confidential and delivery-guaranteed, is formalised by a Notes event of the form 12.2 discussed above (§12.2.2). Rule *Fake1st* lets the Spy open a secure channel to TTP and send a fake message. Rule *AGHP1* represents the first protocol message. In rule *AGHP2*, a Notes event represents R's message to TTP; here, function RPwd indicates the receiver's password. Because messages 2 and 3 travel over guaranteed delivery channels, the protocol model does not require a rule of the form of *Reception1st* (§12.2.2). Hence, the subjects of the Notes events in rules *AGHP2* and *AGHP3* respectively are the intended recipients of the messages (TTP and R respectively).

Steps 3 and 4 must take place at the same time, so they are formalised by the single rule *AGHP3*. TTP checks R's password to authenticate the sender of message 2, but, regardless, he must reply along the same secure channel. The replies to both S and R are delivery-guaranteed, so the rule introduces an appropriate Notes event for the receiver, and a double Says-Gets event for TTP's transmission to the sender. The Says event may seem unnecessary, but it preserves a feature of our model: every Gets event has a matching Says event.

13.2.2 Verification

We focus on the accountability properties, omitting the classical ones on confidentiality and authentication, which conform to the numerous examples seen so far. Likewise, obvious regularity lemmas hold for the private signature or encryption keys of uncompromised agents.

The guarantees presented here supersede those published some time ago [35]. They together confirm the main goal of the protocol, that the sender S gets the return receipt if and only if the receiver R gets the e-mail. Here, *evs* is a generic trace of the formal protocol model aghp. As with the non-repudiation protocol, "$X \in$ used *evs*" is worded as "X exists on *evs*."

Proving validity of evidence. This protocol offers no non-repudiation of origin to the receiver. The main guarantee of validity of S's evidence says that if the return receipt exists then R has obtained the cryptographic key necessary to retrieve the e-mail (Theorem 13.2.1). This guarantee is for S: he can check the form of *S2TTP* as he created it.

Theorem 13.2.1 (AGHP_RR_validity). *If R is not the Spy, RR exists on evs and*

$$RR = \mathsf{Crypt}(\mathsf{priSK\ TTP})\ S2TTP$$

and

$$S2TTP = \mathsf{Crypt}(\mathsf{pubEK\ TTP})\{\!|\mathsf{Agent}\ S, \mathsf{Number}\ AO, \mathsf{Key}\ K, \mathsf{Agent}\ R, hs|\!\}$$

and

$$hs = \mathsf{Hash}\{\!|\mathsf{Number}\ cleartext, \mathsf{Nonce}\ q, r, c|\!\}$$

then evs contains

Nil :
[] ∈ aghp

Fake :
⟦ evsF ∈ aghp; X ∈ synth(analz(knows Spy evsF)) ⟧
⟹ Says Spy B X # evsF ∈ aghp

Fake1st :
⟦ evsF1 ∈ aghp; X ∈ synth(analz(knows Spy evsF1)) ⟧
⟹ Notes TTP ⦃Agent Spy, Agent TTP, X⦄ # evsF1 ∈ aghp

Reception :
⟦ evsR ∈ aghp; Says A B X ∈ set evsR ⟧ ⟹ Gets B X # evsR ∈ aghp

AGHP1 :
⟦ evs1 ∈ aghp; Key K ∉ used evs1; K ∈ symKeys;
  Nonce q ∉ used evs1;
  hs = Hash ⦃Number cleartext, Nonce q, response S R q,
            Crypt K (Number m)⦄;
  S2TTP = Crypt (pubEK TTP) ⦃Agent S, Number BothAuth, Key K,
                            Agent R, hs⦄ ⟧
⟹ Says S R ⦃Agent S, Agent TTP, Crypt K (Number m),
            Number BothAuth, Number cleartext, Nonce q, S2TTP⦄
      # evs1 ∈ aghp

AGHP2 :
⟦ evs2 ∈ aghp; TTP ≠ R;
  Gets R ⦃Agent S, Agent TTP, c, Number BothAuth,
          Number cleartext, Nonce q, S2TTP⦄ ∈ set evs2;
  hr = Hash ⦃Number cleartext, Nonce q, response S R q, c⦄ ⟧
⟹ Notes TTP ⦃Agent R, Agent TTP, S2TTP, Key(RPwd R), hr⦄
      # evs2 ∈ aghp

AGHP3 :
⟦ evs3 ∈ aghp; TTP ≠ R; hs = hr; K ∈ symKeys;
  Notes TTP ⦃Agent R, Agent TTP, S2TTP, Key(RPwd R), hr⦄
    ∈ set evs3;
  S2TTP = Crypt (pubEK TTP) ⦃Agent S, Number BothAuth, Key K,
                            Agent R, hs⦄ ⟧
⟹ Notes R ⦃Agent TTP, Agent R, Key K, hr⦄ #
    Gets S (Crypt (priSK TTP) S2TTP) #
    Says TTP S (Crypt (priSK TTP) S2TTP)
      # evs3 ∈ aghp

Fig. 13.5. Inductive model of the certified e-mail protocol

Notes R {|Agent TTP, Agent R, Key K, hr|}.

The inductive proof is lengthy (eleven commands), with separate consideration of four cases of the induction. Nothing inherently difficult is involved; the complicated form of the assertion causes Isabelle's automatic provers to require more guidance than usual.

Proving fairness. Theorem 13.2.1 seen above expresses fairness for the receiver. Its interpretation in terms of goal availability is subtle, as the receiver can verify the form of $S2TTP$ only in the case of disputes. However, fairness can be equally considered available to him: if the receiver conjectures that the return receipt binding him exists, then he is entitled to obtain the corresponding decryption key. As seen in the previous chapter, a fairness guarantee for an agent must necessarily insist on the very piece of evidence meant for the agent's peer (which often only the peer can verify) and introduce the other piece of evidence meant for the agent.

Before proceeding to the fairness guarantee for the sender, we need to introduce a lemma. It concerns $S2TTP$, the certificate that the sender transmits to TTP in the first protocol message. The lemma says that anything matching the form of $S2TTP$ can only arise from a valid instance of the first protocol message, provided it carries a confidential session key K (Lemma 13.2.1). This assumption can be conventionally relaxed by the appropriate confidentiality argument, which we omitted.

Lemma 13.2.1 (AGHP_S2TTP_sender). *If evs is such that*

Key $K \notin$ analz(knows Spy evs)

and S2TTP exists on evs and

$S2TTP =$ Crypt(pubEK TTP){|Agent S, Number AO, Key K, Agent R, hs|}

then, for some m, ctxt and q, evs contains

Says $S R$ {|Agent S, Agent TTP, Crypt K(Number m), Number AO, Number $ctxt$, Nonce q, $S2TTP$|}

and

$hs =$ Hash{|Number $ctxt$, Nonce q, response $S R q$, Crypt K(Number m)|}.

The proof is straightforward: the Spy needs to know the session key K before he can use it to make a fake version of $S2TTP$. However, once again the proof script requires some guidance for Isabelle's provers.

The fairness guarantee for the sender is expressed as two theorems: one for when the receiver is the Spy and one for an honest receiver. The sender does not have to know which case applies, but only needs to be able to check the theorem assumptions in practice.

In the former case (Theorem 13.2.2), the session key clearly is not confidential. The main premise is that the sender has issued message 1 (with the given value of $S2TTP$). The conclusion is that the receiver is compromised; but even in this case, the sender gets the return receipt.

Theorem 13.2.2 (AGHP_S_fairness_bad_R). *If S is not the Spy and*

$S2TTP = \mathsf{Crypt}(\mathsf{pubEK\,TTP})\{\!|\mathsf{Agent}\,S, \mathsf{Number}\,AO, \mathsf{Key}\,K, \mathsf{Agent}\,R, hs|\!\}$

and

$RR = \mathsf{Crypt}(\mathsf{priSK\,TTP})\,S2TTP$

and evs contains

Says $S\,R$ $\{\!|\mathsf{Agent}\,S, \mathsf{Agent\,TTP}, \mathsf{Crypt}\,K(\mathsf{Number}\,m), \mathsf{Number}\,AO,$
$\qquad\qquad \mathsf{Number}\,cleartext, \mathsf{Nonce}\,q, S2TTP|\!\}$

and is such that

Key $K \notin \mathsf{analz}(\mathsf{knows\,Spy}\,evs)$

then R is compromised and evs contains

Gets $S\,RR$.

The proof script is a simple induction except for the treatment of the third protocol message, when TTP replies to the sender and to the receiver. The proof of this subgoal is rather subtle, resembling a confidentiality argument. Any assertion of the form "if the Spy knows K then ..." is a confidentiality property. In particular, if the conclusion simply is "False" then the assertion is equivalent to saying that the key is confidential. Here, we get a case analysis on whether the Spy knows the key or not, and in the latter case we appeal to Lemma 13.2.1 and to a unicity guarantee (**AGHP_Key_unique**, omitted here). Even with this complicated argument, Isabelle's provers do much of the work, and the treatment of the third message consists of only five commands.

The other case of the fairness argument for the sender is when the receiver is uncompromised. This time, the receiver legitimately gets the session key to decrypt the e-mail. The conclusion is unvaried, that the sender gets the return receipt (Theorem 13.2.3).

Theorem 13.2.3 (AGHP_S_guarantee). *If S is not the Spy and*

$S2TTP = \mathsf{Crypt}(\mathsf{pubEK\,TTP})\{\!|\mathsf{Agent}\,S, \mathsf{Number}\,AO, \mathsf{Key}\,K, \mathsf{Agent}\,R, hs|\!\}$

and

$RR = \mathsf{Crypt}(\mathsf{priSK\,TTP})\,S2TTP$

and evs contains

Says $S\,R$ $\{\!|\mathsf{Agent}\,S, \mathsf{Agent\,TTP}, \mathsf{Crypt}\,K(\mathsf{Number}\,m), \mathsf{Number}\,AO,$
$\qquad\qquad \mathsf{Number}\,cleartext, \mathsf{Nonce}\,q, S2TTP|\!\}$ *and*
Notes R $\{\!|\mathsf{Agent\,TTP}, \mathsf{Agent}\,R, \mathsf{Key}\,K, hs|\!\}$

then evs contains

Gets S RR.

In this case, the proof script is surprisingly short. The argument for the crucial third message consists of a single prover call to `blast`, invoking four available results. Two are Lemma 13.2.1 and Theorem 13.2.2. Another says that whatever is sent on an SSL channel (**ZG_Notes_SSL_imp_used**, omitted here) is a used component. Finally, a unicity theorem is needed about message 1, established using the fresh session key as a pivot (**ZG_Key_unique**, omitted here). Remarkably, Isabelle has to cope with a rather intricate proof tree.

Also this fairness guarantee must be interpreted with care in terms of goal availability. The Notes R fact is, rather than an assumption that S must be able to verify, the necessary conjecture that S makes to conclude that he would get his own piece of evidence. In consequence, fairness is available to the sender.

We have emphasised throughout that each proof about the certified e-mail protocol relies on assumptions that the beneficiary of the corresponding theorem can check. So, it can be concluded that the protocol makes its goals available to its peers. However the proof development process highlighted that an anomalous execution of the protocol is possible. The receiver can initiate a session from step 2 by quoting an arbitrary sender, and by building two identical hashes. The session will terminate successfully and the sender will get evidence that an e-mail he has never sent has been delivered. This is due to the fact that the protocol uses no technique to authenticate the sender to TTP. The anomaly can be addressed by inserting the sender's password into the certificate $S2TTP$ created at step 1, so that the receiver cannot forge it. But passwords are weak secrets.

Another flaw is that the sender has no defence against the receiver's claim that the message was sent years ago and is no longer relevant, which would devalue the return receipt. This attack works in both directions: the receiver's claim might be truthful and not believed. Even if the sender includes a date in the message, he cannot prove that the date is accurate. The obvious solution is for TTP to include a timestamp in the return receipt.

13.3 Discussion

There exist two pen-and-paper analyses of Zhou and Gollman's protocol. One, by the designers themselves [168], uses the SVO authentication logic; the other, by Schneider [145], uses rank functions and CSP. An automated analysis by Gürgens and Rudolph [87] has found replay attacks under the assumption that TTP does not maintain an audit trail. None of these analyses was directly useful to us, as our approaches appear to be rather dissimilar. As for the protocol by Abadi et al., our findings appear to be coherent with those by Abadi and Blanchet on the same protocol [2], which were developed using an *ad hoc* proof tool, ProVerif [51]. Shmatikov and Mitchell use a finite-state checker for protocols that are similar to ours [147].

We have developed simple formalisations of non-repudiation and certified e-mail delivery, along with simple proof methods relying on induction. The formalisation of a distrusted peer differs in the two protocols. With the non-repudiation protocol we had to assume the peer to be unbroken because that protocol is based on digital signatures, which are worthless if the peer's private keys have been disclosed. Both protocols make their goals available to their peers, but the certified e-mail protocol is based upon weaker mechanisms: passwords and previously agreed responses. Although this was an explicit choice of the designers, it must be kept in mind when interpreting the theorems proved in our model. Guarantees based on strong cryptography are sounder than those based on weak passwords.

Our findings confirm that both the non-repudiation protocol and the certified e-mail protocol achieve their stated goals. All evidence appears to be valid: sufficient to hold an agent accountable for participation. The non-repudiation protocol delivers evidence to its participants, binding each other's participation. It is fair: each party receives evidence if and only if the other party does. The certified e-mail protocol is fair in the sense that the initiator gets non-repudiation evidence — the return receipt — if and only if the responder gets the e-mail. Interpreting fairness in terms of goal availability was subtler with the latter protocol. Evidence in the non-repudiation protocol consists of digital signatures that anyone can verify upon reception. By contrast, evidence in the certified e-mail protocol comprises the session key used to build the commitment and the certificate $S2TTP$ encrypted with TTP's public key. Assumptions on such message components must be interpreted as conjectures that an agent makes to find out whether he would get what he seeks should those conjectures hold.

Comparing the two protocols, the e-mail one demands less of the trusted third party and it uses much weaker cryptographic mechanisms, with no public-key infrastructure. It offers correspondingly weaker guarantees: the responder gets no non-repudiation evidence, and even the theorems we can prove must be interpreted with an awareness of the weak cryptography.

A reviewer of one of our earlier papers suggested that it is interesting to see what happens if a protocol is deliberately weakened. For example, suppose that the receiver of the certified e-mail protocol forgets to send message 2 over the SSL protocol and sends it over a conventional transport protocol. Modelling this variant is straightforward: replace the event

Notes TTP {|Agent R, Agent TTP, $S2TTP$, Key(RPwd R), hr|}

in rule $AGHP2$ with the event

Says R TTP {|$S2TTP$, Key(RPwd R), hr|}

and fix the corresponding premise in rule *AGHP3* accordingly. An attempt to reexecute our proof script soon reveals that the receiver is sending his secret password in the clear and disclosing it to the Spy. Specifically, the regularity lemma stating that the Spy only knows the passwords of compromised agents (**AGHP_Spy_analz_RPwd**, omitted here) fails in the new model. This kind of experiment may seem interesting, but it is unnecessary. The Inductive Method works by establishing facts through formal proof of correctness. The chain of reasoning is open to inspection. This is fundamentally different from model checkers and other automatic analysis tools.

Our methods scale up to analysing accountability protocols. We have examined two protocols that have similar goals but operate in very different security environments. Numerous, though straightforward, changes were necessary to model the novel architectures.

14. Conclusions

Establishing secure communication sessions with remote computers is vital in our era of computer networks. Security stands for a variety of goals, which may vary depending on the application domain. Confidentiality appears to remain the best known and understood concept. But more recent properties, such as various forms of agent accountability, are deepening their impact. Enforcing those goals can be daunting, as it requires a deep understanding of their significance and especially a clear connotation of the threats they are supposed to withstand.

Practical experience has confuted many informal claims of security and some formal ones. The e-commerce market is seriously concerned that most purchase attempts terminate exactly at the stage of entering the credit card details. Security is often ill understood by researchers, let alone by the general population. Recently, a number of engineers without specific training in security reduced security to mere cryptography straight away during a private conversation. People in a cafeteria easily talked about connecting to the Internet, but with difficulty about security issues. Here is their conversation reported verbatim.

- A: *"I surf the Internet just fine!"*
- B: *"I can't! My computer says it's got some ports closed!"*
- A: *"Then it must be a virus!"*
- B: *"Doesn't a virus open up your ports instead?"*

These experiences may sound trivial. However, they could be emblematic of our times: we might just be witnessing the very beginning of the era of *secure computer networks*. Our book is born in this atmosphere. It shows us how to reason about the smallest subtleties of security protocols, thus contributing to the general understanding of a core problem in the era of secure computer networks. It also teaches us a specific formal method of protocol analysis, and basic interaction with the theorem prover Isabelle. The main conclusion is that the protocol verification problem can be effectively tackled by formal methods with the aim of developing security protocols that indeed are secure.

Security protocols are an important piece in the security puzzle. Understanding them thoroughly in turn means understanding a number of security issues. While an informal language can favour human intuition, a formal one can help unveil secluded features. The Inductive Method (Chapter 3) in essence advances an approach to protocol analysis based on mathematical proof of correctness. Each Isabelle proof can be easily inspected by humans, who get the chance to decide whether the proof can be considered valid. This might contribute to the evolution of secure computer networks better than any boolean answers provided by mysterious press-button tools.

Various approaches have been taken to analysing security protocols formally (Chapter 2). The earlier ones used known techniques of abstract logics that had been established in other contexts, and tailored them to the new application. Various techniques have been developed for the explicit purpose of protocol analysis, such as strand spaces or special-purpose theorem provers. For their own specialised nature, these seem to obtain some of the most significant findings. However, established general-purpose techniques, such as CSP used with model checking, have rarely failed to provide significant contributions. It is difficult to choose the best approach, or perhaps impossible. Notoriously, a group of methods used in conjunction may achieve the best results in practice. However, with increasingly more complex protocols being developed on top of existing ones, some support for composition of existing results might be decisive in the future.

Composition of proofs is not currently supported in the Inductive Method. Blanqui's protocol-independent secrecy [33, 34] is significant though not exactly in this vein. It rather counts as an attempt to generalise the formal reasoning regardless of the specific protocol under analysis. Our brief account of second-level protocols (Chapter 12) is a simplification of actual proof composition. The goals of the underlying protocols become available assumptions to proving the goals of the protocols obtained by composition. It is not obvious that a proof obtained by composition would hold for the correspondingly composed protocol: the proof should also account for hidden interactions between the protocols, which are not an issue in a stand-alone context [10, 66].

Formal analysis in general has at least one other merit: it keeps the focus on the assumed threats, which, history shows, are often neglected. Along these lines, our goal availability principle (Chapter 5) stresses that explicitness deserves no consideration without a detailed account of the underlying threat model. Goal availability by itself stresses the importance of conducting formal reasoning from the agents' viewpoints. This will not simply result in expressive formal statements but, more importantly, in new protocol insights.

If the goals of security protocols are not trivial to understand [81], then they are less easy to study formally, and the resulting guarantees may be difficult to interpret (Chapter 2). When our research began in late 1996, the following goals had already been treated in the Inductive Method (Chapter 4): regularity, unicity, confidentiality and authentication. We have substantially deepened the treatment of the protocol goals as follows.

- A variety of subsidiary guarantees have been interpreted as reliability of the formal protocol model.
- The regularity properties have been extended to the context of smartcard protocols, which involve additional long-term keys.
- A number of existing guarantees have been reinterpreted in terms of authenticity of messages, and their relation with integrity guarantees clarified.
- Novel unicity properties have been designed stating that an event can only occur once (Chapter 6), or relying on fresh timestamps (Chapter 9).
- A stronger version of authentication called non-injective agreement has been studied formally.
- The goal of key distribution has been treated formally.
- Accountability goals of evidence validity and fairness have received formal treatment in the context of non-repudiation and certified e-mail (Chapter 12).

Confidentiality was the only goal that would not require substantial extensions if not for the generalisation, required by Kerberos IV, of a former proof method. By contrast, the novel goals typically required the development of additional proof methods. This can also be observed from our recent analysis of the SET protocol with Massacci and Paulson [36, 37, 38], not part of this book, which, for example, adopts our formalisation of message reception (Chapter 8).

The formal treatment of timestamping turned out to be reasonably simple, although it required some formalisation of the concept of time. It was used to analyse three versions of the Kerberos protocol: BAN (Chapter 6), IV (Chapter 7) and V (Chapter 9). It also allowed for a temporal modelling of accidents on session keys. In particular, Kerberos IV required moderate extensions to existing proof methods for confidentiality, and Kerberos V unveiled alternative and faster proof methods for authentication, due precisely to its main feature of disposing with the previous version's use of double encryption.

To complete a summary of this book's contribution, the treatment of agents' knowledge must be recalled (Chapter 8). It developed around two options, one in terms of mere message creation (trace inspection) and the other in terms of message deducibility (message sending/receiving), and was variously demonstrated throughout. It was fundamental to studying the goals of non-injective agreement and key distribution. Another extension was an account of smartcards, with associated long-term keys, computational power and risks of loss or cloning (Chapter 10).

The Shoup-Rubin protocol was then studied in a realistic threat model for its smartcards (Chapter 11) and some lack of explicitness was highlighted. Finally the non-repudiation protocol by Zhou and Gollmann was comparatively analysed with the certified e-mail protocol by Abadi et al. (Chapter 13), thanks to the dedicated extensions previously introduced in the Inductive Method (Chapter 12).

The future of protocol verification is multifaceted. On the one hand, credit card insurances currently seem the most common way to face electronic fraud. On the other hand, fraud prevention seems ideal, so increasingly more research in protocol correctness will become necessary to cope with emerging goals. However, not all security properties become important goals. To give just a few examples, the last years of the 1990s saw widespread rumours about *anonymity*, which seems of interest only in very restricted contexts at present. Conversely, *non-denial of service* has become a stringent need. A technique to counter denial of service is *cookie transformation* [69]: move the computational load onto the client, so that a server will engage in a session only if the client previously did substantial computation. Very few attempts to study this problem formally exist at present [114]. It seems fair to state that the Inductive Method has reached a level of maturity to easily cope with most emerging goals, even with protocols aiming at cookie transformation. We can envisage some formalisation of this property in terms of a safety guarantee, as it remains clear that no liveness guarantee (namely that something *will* happen) can be proved by induction.

Another line of research concerns the continuation of the analysis after an attack is found [32]. It involves assuming that any agent can be the Spy, acting for his own sake. This extended threat model is similar to the one we used to analyse the accountability protocols, and seems perfectly plausible in the present technical/social setting. It prepares the ground for unexplored properties such as *retaliation*: if Alice attacks Bob, then Bob can attack Alice back. Although easy to understand, such a goal is not trivial to formalise as it may involve trace properties with at least double quantification. Proving retaliation might have important social consequences. An attacker might opt for refraining from attacking if the risk of being attacked back is considerable. It is expected that retaliation goals can be analysed inductively although the new threat model will increase the computational burden.

Before concluding with some statistics about the findings discussed in this book, we remind the reader that all our proofs of correctness are available from the 2006 distribution of Isabelle [33, 34] (before that distribution is released, they can be found with the development snapshot [156]). It means that they will be maintained and will remain fully functional while Isabelle develops throughout the years, and hence will be available for researchers and practitioners' inspection also with future distributions of Isabelle.

Inspection can help appreciate in depth the entangled design strategies of a number of protocols, some classical, others already deployed at the beginning of this decade. As those design strategies are most likely to survive for decades, we believe and, at the same time, hope that the results discussed in this book will also serve to develop correct security protocols in the future.

14.1 Statistics

There is concern that mathematical proofs are getting harder to verify by humans: "A mathematical proof is irrefutably true, a manifestation of pure logic. But an increasing number of mathematical proofs are now impossible to verify with absolute certainty" [100] as they often span several thousand lines. However, this does not apply to our proofs; some statistics confirm that the scripts are of a manageable size. It is hard to express the human effort necessary to develop the proofs, and statistics can only give a macroscopic perspective.

14.1.1 Theory File Sizes

Some readers may find the sheer size of the theory files indicative of their level of complication. Table 14.1 shows the total number of lines in each file. That includes the specification of the protocol and necessary functions as well as the theorem statements and their proof scripts. To give some idea of the actual proof complications, the table also presents the number of proof methods applied in each file.

Theory file	Total lines	Proof methods
Message.thy	969	216
Event.thy	357	51
Public.thy	451	77
NS_Shared.thy	528	112
Kerberos_BAN.thy	729	161
Kerberos_BAN_Gets.thy	720	145
KerberosIV.thy	1925	516
KerberosIV_Gets.thy	1547	397
KerberosV.thy	1654	449
OtwayReesBella.thy	392	80
ZhouGollmann.thy	463	95
CertifiedEmail.thy	489	113
EventSC.thy	447	96
Smartcard.thy	474	70
ShoupRubin.thy	1395	206
ShoupRubinBella.thy	1389	212
Total	**13929**	**2996**

Table 14.1. Theory file sizes

As a rule of thumb, the larger the total number of lines, the larger the protocol both in terms of specification and in terms of relevant properties. Likewise, the larger the number of proof methods, the more complicated the proofs. For example, a smartcard protocol counts approximately 1400 lines in total, which is not too far from some Kerberos theory files, such as that for version IV with message reception. However, a smartcard protocol only features some 200 applications of proof methods, namely half as many as the mentioned Kerberos version does. Such a simple comparison signifies that Kerberos is shorter to specify but longer to prove correct.

If we consider the various Kerberos versions, we find that the BAN version is smaller in both senses than the other versions, as can be expected. Somewhat surprisingly, version IV formalised with message reception seems smaller than the corresponding one without reception: KerberosIV_Gets.thy has only 1547 lines. But this is not meaningful as the smaller version purposely omits a few minor theorems. The corresponding values for the BAN version, which remain uniform after message reception, are more significant.

It is visible that Smartcard.thy resembles, and slightly extends, Public.thy as it logically lies (Figure 3.1) at the same level: 474 total lines versus 451. But the method applications negligibly decrease from 77 to 70. A similar comparison reveals that the theory file of events for smartcards, EventSC.thy, is appreciably more complicated than the original theory file: 474 total lines versus 357, and 70 applications of proof methods versus 51.

Two other theory files are interesting to compare: NS_Shared.thy and OtwayReesBella.thy. The first contains the entire proof script for the protocol, while the second only features the fragment that is relevant to proving the key distribution goal. But both values decrease proportionally from the first to the second file. Finally, the two accountability protocols present coherent values. Theory CertifiedEmail.thy has only 19% more proof method applications than theory ZhouGollmann.thy.

The entire proof script counts some 3000 applications of proof methods and nearly 14000 total lines!

14.1.2 Proof Runtimes

Another important indicator of the complexity of proofs is how long they take to execute. We measured the runtimes on two currently inexpensive configurations, one running Linux and Poly/ML and the other running Cygwin on Windows and SML/NJ. Here are their details.

Configuration A

- AMD Athlon XP 1500GHz
- 512 MB RAM
- Slackware 10.2, Linux kernel 2.6.14.3
- Poly/ML 4.1.4
- Isabelle development snapshot 13 February 2006
- Proof General 3.6pre050930

Configuration B

- Intel Pentium M 760 (2GHz)
- 1 GB RAM
- Microsoft Windows XP Service Pack 2 and Cygwin 1.5.19-4
- SML/NJ 110.49
- Isabelle development snapshot 13 February 2006
- Proof General 3.6pre050930

Clearly, configuration B is more powerful than configuration A at least in terms of hardware, so we expect a better performance despite the Linux emulation offered by Cygwin. Our measurements given in Table 14.2 show that configuration B nearly halves the performance of configuration A: 522 seconds versus 917.

Theory file	Configuration A	Configuration B
Message.thy	12	7
Event.thy	12	6
Public.thy	14	8
NS_Shared.thy	40	23
Kerberos_BAN.thy	36	20
Kerberos_BAN_Gets.thy	42	23
KerberosIV.thy	140	83
KerberosIV_Gets.thy	150	85
KerberosV.thy	138	82
OtwayReesBella.thy	30	17
ZhouGollmann.thy	32	17
CertifiedEmail.thy	68	40
EventSC.thy	20	11
Smartcard.thy	11	6
ShoupRubin.thy	83	45
ShoupRubinBella.thy	89	49
Total	**917**	**522**

Table 14.2. Proof runtimes in seconds

It is easy to observe that the heaviest theory file is KerberosIV_Gets.thy: it executes in 150 seconds on configuration A and 85 seconds on configuration B, which is approximately 44% faster. Similar theory files, also for Kerberos IV, produce comparable runtimes.

The files for the smartcard protocols execute in less than 90 seconds and less than 50 seconds in the two configurations respectively. These are about 58% of the mentioned Kerberos runtimes. Although the smartcard protocols are longer to specify and perhaps understand, Kerberos IV and V are more complex to reason about. This was also confirmed above by the much larger number of applications of proof methods. It is the confidentiality argument that is especially more complex and longer in Kerberos, due to its pair of session keys. It follows that reasoning about the analz operator is particularly time consuming.

The runtimes of the basic theory files of messages, events and smartcards are not surprising. Neither are those for NS_Shared.thy and Kerberos_BAN.thy. What was not expected was that the runtime of the certified e-mail protocol more than doubles that of the non-repudiation protocol, although the former has only 19% more proof methods, as mentioned above. This finding inspires a closer comparison of the two proof scripts. It appears that, on average, proving a subgoal of the certified e-mail protocol requires a deeper analysis to specifically deal with the underlying goals of first-level security protocols. It raises counterintuitive concerns that a hierarchical style of verification might be more time consuming than a flat style, but further investigations are necessary before turning those concerns into worries.

14.1.3 Human Effort

Measuring the human effort is especially difficult when the researchers do not work full-time on a single project. We stated that our findings span over nearly a decade. However, especially after the first four years (of the author's Ph.D. course), effort was only spent part-time. Nonetheless, some approximations are provided in Table 14.3.

Theory file	Man weeks (MW)	Difficulty % (DP)
Message.thy	/	/
Event.thy	1	23
Public.thy	/	/
NS_Shared.thy	1	10
Kerberos_BAN.thy	12	75
Kerberos_BAN_Gets.thy	8	70
KerberosIV.thy	22	90
KerberosIV_Gets.thy	1	20
KerberosV.thy	3	55
OtwayReesBella.thy	5	65
ZhouGollmann.thy	10	80
CertifiedEmail.thy	12	80
EventSC.thy	14	70
Smartcard.thy	8	60
ShoupRubin.thy	16	60
ShoupRubinBella.thy	2	25
Total	114	/

Table 14.3. Approximated human efforts

The approximations are expressed in terms of man-weeks (MW) of the author's time, and enriched with a difficulty percentage (DP), which is entirely subjective. No values are available for the original theory files `Message.thy` and `Public.thy`, already existing before our work. The moderate indicator values for `Event.thy` reflect the updates to formalise message reception.

The two indicators are in general not trivially related. For example, `Kerberos_BAN.thy` was developed in approximately three months (MW=12) and with significant difficulty (DP=75) due to the formalisation of timestamping. The corresponding version enriched with the reception event was one third quicker to develop, although its difficulty percentage remains as high as 70. This reflects the first experiments with the reception event, which was quicker and easier to incorporate in subsequent analyses.

The statistics about Kerberos IV are significant. It can be seen that `KerberosIV.thy` was the longest (MW=22) and hardest theory file (DP=90) to develop. This is due to the difficult confidentiality goals and to the first account of agents' knowledge on the basis of trace inspection. Interestingly, the corresponding version enriched with the reception event was extremely quick (MW=1) thanks to previous experience with the BAN version, but not proportionally trivial (DP=20) due to the delicate updates to some subsidiary lemmas. Kerberos V was finished rather quickly (MW=3), but the necessity of novel proof methods increased its difficulty percentage (DP=55).

The indicator values for `NS_Shared.thy` only refer to the proofs of the Issues properties, and hence are rather low, as the main properties of this protocol had already been studied by Paulson. Also those about `OtwayReesBella.thy` only refer to the innovative proof of the key distribution goal, but the indicator values (MW=5, DP=65) are emblematic of the originality of the underlying ideas.

It can be seen that the indicator values for the two accountability protocols are almost identical. The slightly longer time for `CertifiedEmail.thy` betrays the various attempts to formalise second-level protocols. We then come to the formalisation of smartcards. The theory file `EventSC.thy` required more than three months (MW=14), and was significantly difficult (DP=70) not only for the introduction of the extra events for smartcards but also, and especially, for the delicate formalisation of agents' knowledge depending on the assumption of secure means. Then, `Smartcard.thy` took less time (MW=8) but was comparatively harder (DP=60) due to some initial confusion with the usability of the cards over secure or insecure means.

Finally, the actual smartcard protocol was rather long to deal with (MW=16) because it is the longest protocol ever analysed inductively. Its formal model is longer than that of the purchase phase of SET [38]. However, it was proportionally easy to deal with (DP=60) as the main proofs are long rather than complicated (one may have many facts to prove, but easily). Our updated version `ShoupRubinBella.thy`, which did not seem to be easy work

(DP=25), especially for the interpretative efforts, was developed in just a couple of weeks (MW=2).

Concluding with the human effort, the preparation of this manuscript took about 34 man-weeks to be added to the approximately 40 for the preparation of the author's Ph.D. thesis on which this book is based. These values do not include the development of general concepts, such as goal availability, which did not involve interaction with the theorem prover: some 20 man-weeks. Therefore, the grand total is 208 man-weeks.

A. Proof Script Fragments for Kerberos IV

.

A.1 Reliability

The following describes the form of all components sent by Kas.

```
lemma Says_Kas_message_form:
    "⟦ Says Kas A
          (Crypt K ⦃Key authK, Agent Peer, Number Ta, authTicket⦄)
          ∈ set evs;
          evs ∈ kerbIV ⟧ ⟹
  K = shrK A  & Peer = Tgs &
  authK ∉ range shrK & authK ∈ authKeys evs & authK ∈ symKeys &
  authTicket =
       (Crypt (shrK Tgs) ⦃Agent A, Agent Tgs, Key authK, Number Ta⦄) &
  Key authK ∉ used(before
    Says Kas A (Crypt K ⦃Key authK, Agent Peer, Number Ta, authTicket⦄)
                 on evs) &
  Ta = CT (before
    Says Kas A (Crypt K ⦃Key authK, Agent Peer, Number Ta, authTicket⦄)
           on evs)"
apply (unfold before_def)
apply (erule rev_mp)
apply (erule kerbIV.induct)
apply (simp_all (no_asm) add: authKeys_def authKeys_insert,
       blast, blast)
```

    K2

```
apply (simp (no_asm) add: takeWhile_tail)
apply (rule conjI)
apply clarify
apply (rule conjI)
apply clarify
apply (rule conjI)
apply blast
apply (rule conjI)
apply clarify
apply (rule conjI)
```

Subcase: used before.

```
apply (blast dest: used_evs_rev [THEN equalityD2,
       THEN contra_subsetD] used_takeWhile_used)
```

Subcase: CT before.

```
apply (fastsimp dest!: set_evs_rev [THEN equalityD2,
       THEN contra_subsetD, THEN takeWhile_void])
apply blast
```

Rest

```
apply blast+
done
```

```
lemma K3_imp_K2:
    "⟦ Says A Tgs
           {|authTicket, Crypt authK {|Agent A, Number T2|}, Agent B|}
           ∈ set evs;
         A ∉ bad;  evs ∈ kerbIV ⟧
    ⟹ ∃Ta. Says Kas A (Crypt (shrK A)
                      {|Key authK, Agent Tgs, Number Ta, authTicket|})
                  ∈ set evs"
apply (erule rev_mp)
apply (erule kerbIV.induct)
apply (frule_tac [7] K5_msg_in_parts_spies)
apply (frule_tac [5] K3_msg_in_parts_spies, simp_all, blast, blast)
apply (blast dest: Says_imp_spies [THEN parts.Inj,
       THEN authK_authentic])
done
```

Describes the form of all compontents sent by Tgs.

```
lemma Says_Tgs_message_form:
    "⟦ Says Tgs A
        (Crypt authK {|Key servK, Agent B, Number Ts, servTicket|})
           ∈ set evs;
         evs ∈ kerbIV ⟧
  ⟹ B ≠ Tgs &
      authK ∉ range shrK & authK ∈ authKeys evs & authK ∈ symKeys &
      servK ∉ range shrK & servK ∉ authKeys evs & servK ∈ symKeys &
      servTicket =
        (Crypt (shrK B) {|Agent A, Agent B, Key servK, Number Ts|}) &
      Key servK ∉ used (before
  Says Tgs A (Crypt authK {|Key servK, Agent B, Number Ts, servTicket|})
                      on evs) &
         Ts = CT(before
  Says Tgs A (Crypt authK {|Key servK, Agent B, Number Ts, servTicket|})
                  on evs) "
apply (unfold before_def)
apply (erule rev_mp)
apply (erule kerbIV.induct)
apply (simp_all add: authKeys_insert authKeys_not_insert
       authKeys_empty authKeys_simp, blast)
```

We need this simplification only for Message 4.

```
apply (simp (no_asm) add: takeWhile_tail)
apply auto
```

Five subcases of Message 4.

```
apply (blast dest!: SesKey_is_session_key)
apply (blast dest: authTicket_crypt_authK)
apply (blast dest!: authKeys_used Says_Kas_message_form)
```

Subcase: used before.

```
apply (blast dest: used_evs_rev [THEN equalityD2, THEN
      contra_subsetD] used_takeWhile_used)
```

Subcase: CT before.

```
apply (fastsimp dest!: set_evs_rev [THEN equalityD2,
      THEN contra_subsetD, THEN takeWhile_void])
done
```

```
lemma authTicket_form:
    "⟦ Crypt (shrK A) ⦃Key authK, Agent Tgs, Ta, authTicket⦄
          ∈ parts (spies evs);
       A ∉ bad;
       evs ∈ kerbIV ⟧
    ⟹ authK ∉ range shrK & authK ∈ symKeys &
    authTicket = Crypt (shrK Tgs) ⦃Agent A, Agent Tgs, Key authK, Ta⦄"
apply (erule rev_mp)
apply (erule kerbIV.induct)
apply (frule_tac [7] K5_msg_in_parts_spies)
apply (frule_tac [5] K3_msg_in_parts_spies, simp_all)
apply (blast+)
done
```

A.2 Session-key Compromise

Big simplification law for session keys that are not encrypted by keys in a given set KK. It helps us prove three, otherwise harder, facts about keys. These facts are exploited as simplification laws for analz, and also "limit the damage" in the case of loss of a key to the Spy.

```
lemma Key_analz_image_Key [rule_format (no_asm)]:
    "evs ∈ kerbIV ⟹
    (∀SK KK. SK ∈ symKeys & KK <= -(range shrK) ⟶
    (∀K ∈ KK. ¬ AKcryptSK K SK evs)    ⟶
    (Key SK ∈ analz (Key'KK Un (spies evs))) =
    (SK ∈ KK | Key SK ∈ analz (spies evs)))"
apply (erule kerbIV.induct)
apply (frule_tac [10] Oops_range_spies2)
apply (frule_tac [9] Oops_range_spies1)
apply (frule_tac [7] Says_tgs_message_form)
apply (frule_tac [5] Says_kas_message_form)
apply (safe del: impI intro!: Key_analz_image_Key_lemma [THEN impI])
```

Case-splits for Oops1 and message 5: the negated case simplifies using the induction hypothesis.

```
apply (case_tac [11] "AKcryptSK authK SK evsO1")
```

```
apply (case_tac [8] "AKcryptSK servK SK evs5")
apply (simp_all del: image_insert
      add: analz_image_freshK_simps AKcryptSK_Says shrK_not_AKcryptSK
            Oops2_not_AKcryptSK Auth_fresh_not_AKcryptSK
        Serv_fresh_not_AKcryptSK Says_Tgs_AKcryptSK Spy_analz_shrK)
```
— Computationally expensive

Fake

apply spy_analz

K2

apply blast

K3

apply blast

K4

apply (blast dest!: authK_not_AKcryptSK)

K5

apply (case_tac "Key servK ∈ analz (spies evs5) ")

If servK is compromised then the result follows directly...

```
apply (simp (no_asm_simp) add: analz_insert_eq Un_upper2
      [THEN analz_mono, THEN subsetD])
```

...therefore servK is uncompromised.

The AKcryptSK servK SK evs5 case leads to a contradiction.

```
apply (blast elim!: servK_not_AKcryptSK [THEN [2] rev_notE]
      del: allE ballE)
```

Another K5 case.

apply blast

Oops1

```
apply simp
apply (blast dest!: AKcryptSK_analz_insert)
done
```

First simplification law for analz: no session keys encrypt authkeys or shared keys.

```
lemma analz_insert_freshK1:
    "⟦ evs ∈ kerbIV;  K ∈ authKeys evs Un range shrK;
      SesKey ∉ range shrK ⟧
      ⟹ (Key K ∈ analz (insert (Key SesKey) (spies evs))) =
          (K = SesKey | Key K ∈ analz (spies evs))"
apply (frule authKeys_are_not_AKcryptSK, assumption)
apply (simp del: image_insert
            add: analz_image_freshK_simps add: Key_analz_image_Key)
```

done

Second simplification law for analz: no servkeys encrypt any other keys.

lemma analz_insert_freshK2:
    "⟦ evs ∈ kerbIV; servK ∉ (authKeys evs); servK ∉ range shrK;
     K ∈ symKeys ⟧
    ⟹ (Key K ∈ analz (insert (Key servK) (spies evs))) =
        (K = servK | Key K ∈ analz (spies evs))"
apply (frule not_authKeys_not_AKcryptSK, assumption, assumption)
apply (simp del: image_insert
            add: analz_image_freshK_simps add: Key_analz_image_Key)
done

Third simplification law for analz: only one authkey encrypts a certain servkey.

lemma analz_insert_freshK3:
 "⟦ AKcryptSK authK servK evs;
    authK' ≠ authK; authK' ∉ range shrK; evs ∈ kerbIV ⟧
    ⟹ (Key servK ∈ analz (insert (Key authK') (spies evs))) =
            (servK = authK' | Key servK ∈ analz (spies evs))"
apply (drule_tac authK' = authK' in not_different_AKcryptSK, blast,
        assumption)
apply (simp del: image_insert
            add: analz_image_freshK_simps add: Key_analz_image_Key)
done

Alternative formulation.

lemma analz_insert_freshK3_bis:
 "⟦ Says Tgs A
        (Crypt authK ⦃Key servK, Agent B, Number Ts, servTicket⦄)
        ∈ set evs;
      authK ≠ authK'; authK' ∉ range shrK; evs ∈ kerbIV ⟧
    ⟹ (Key servK ∈ analz (insert (Key authK') (spies evs))) =
            (servK = authK' | Key servK ∈ analz (spies evs))"
apply (frule AKcryptSKI, assumption)
apply (simp add: analz_insert_freshK3)
done

A.3 Session-key Confidentiality

If Spy sees the authkey sent in msg K2, then the key has expired.

lemma Confidentiality_Kas_lemma [rule_format]:
     "⟦ authK ∈ symKeys; A ∉ bad; evs ∈ kerbIV ⟧
     ⟹ Says Kas A
                (Crypt (shrK A)
                    ⦃Key authK, Agent Tgs, Number Ta,
        Crypt (shrK Tgs) ⦃Agent A, Agent Tgs, Key authK, Number Ta⦄⦄)
            ∈ set evs ⟶

```
            Key authK ∈ analz (spies evs) ⟶
            expiredAK Ta evs"
apply (erule kerbIV.induct)
apply (frule_tac [10] Oops_range_spies2)
apply (frule_tac [9] Oops_range_spies1)
apply (frule_tac [7] Says_tgs_message_form)
apply (frule_tac [5] Says_kas_message_form)
apply (safe del: impI conjI impCE)
apply (simp_all (no_asm_simp) add: Says_Kas_message_form
      less_SucI analz_insert_eq
      not_parts_not_analz analz_insert_freshK1 pushes)
```

Fake

```
apply spy_analz
```

K2

```
apply blast
```

K4

```
apply blast
```

K5

```
apply (blast dest: servK_notin_authKeysD
      Says_Kas_message_form intro: less_SucI)
```

Oops1

```
apply (blast dest!: unique_authKeys intro: less_SucI)
```

Oops2

```
apply (blast dest: Says_Tgs_message_form Says_Kas_message_form)
done
```

```
lemma Confidentiality_Kas:
    "⟦ Says Kas A
        (Crypt Ka ⦃Key authK, Agent Tgs, Number Ta, authTicket⦄)
          ∈ set evs;
        ¬ expiredAK Ta evs;
        A ∉ bad;  evs ∈ kerbIV ⟧
    ⟹ Key authK ∉ analz (spies evs)"
by (blast dest: Says_Kas_message_form Confidentiality_Kas_lemma)
```

If Spy sees the servkey sent in msg K4, then the key has expired.

```
lemma Confidentiality_lemma [rule_format]:
    "⟦ Says Tgs A
            (Crypt authK
                ⦃Key servK, Agent B, Number Ts,
          Crypt (shrK B) ⦃Agent A, Agent B, Key servK, Number Ts⦄⦄)
            ∈ set evs;
        Key authK ∉ analz (spies evs);
        servK ∈ symKeys;
        A ∉ bad;  B ∉ bad;  evs ∈ kerbIV ⟧
```

```
        ⟹ Key servK ∈ analz (spies evs) ⟶
            expiredSK Ts evs"
apply (erule rev_mp)
apply (erule rev_mp)
apply (erule kerbIV.induct)
apply (rule_tac [9] impI)+
```

— The Oops1 case is unusual: must simplify Authkey ∉ analz (knows Spy (ev # evs)), not letting analz_mono_contra weaken it to Authkey ∉ analz (knows Spy evs), for we then conclude authK ≠ authKa.

```
apply analz_mono_contra
apply (frule_tac [10] Oops_range_spies2)
apply (frule_tac [9] Oops_range_spies1)
apply (frule_tac [7] Says_tgs_message_form)
apply (frule_tac [5] Says_kas_message_form)
apply (safe del: impI conjI impCE)
apply (simp_all add: less_SucI new_keys_not_analzd
        Says_Kas_message_form Says_Tgs_message_form analz_insert_eq
        not_parts_not_analz analz_insert_freshK1
        analz_insert_freshK2 analz_insert_freshK3_bis pushes)
```

  Fake

```
apply spy_analz
```

  K2

```
apply (blast intro: parts_insertI less_SucI)
```

  K4

```
apply (blast dest: authTicket_authentic Confidentiality_Kas)
```

  Oops2

```
  prefer 3
  apply (blast dest: Says_imp_spies [THEN parts.Inj]
      Key_unique_SesKey intro: less_SucI)
```

  Oops1

```
  prefer 2
apply (blast dest: Says_Kas_message_form Says_Tgs_message_form intro:
      less_SucI)
```

  K5. Not obvious how this step could be integrated with the main simplification step. Done in KerberosV.thy

```
apply clarify
apply (erule_tac V = "Says Aa Tgs ?X ∈ set ?evs" in thin_rl)
apply (frule Says_imp_spies [THEN parts.Inj,
      THEN servK_notin_authKeysD])
apply (assumption, blast, assumption)
apply (simp add: analz_insert_freshK2)
apply (blast dest: Says_imp_spies [THEN parts.Inj] Key_unique_SesKey
      intro: less_SucI)
done
```

In the real world Tgs can't check whether an authkey is secure!

```
lemma Confidentiality_Tgs:
    "[ Says Tgs A
        (Crypt authK {|Key servK, Agent B, Number Ts, servTicket|})
            ∈ set evs;
        Key authK ∉ analz (spies evs);
        ¬ expiredSK Ts evs;
        A ∉ bad;  B ∉ bad; evs ∈ kerbIV ]
    ⟹ Key servK ∉ analz (spies evs)"
apply (blast dest: Says_Tgs_message_form Confidentiality_lemma)
done
```

In the real world Tgs CAN check what Kas sends!

```
lemma Confidentiality_Tgs_bis:
    "[ Says Kas A
        (Crypt Ka {|Key authK, Agent Tgs, Number Ta, authTicket|})
            ∈ set evs;
        Says Tgs A
    (Crypt authK {|Key servK, Agent B, Number Ts, servTicket|})
            ∈ set evs;
        ¬ expiredAK Ta evs; ¬ expiredSK Ts evs;
        A ∉ bad;  B ∉ bad; evs ∈ kerbIV ]
    ⟹ Key servK ∉ analz (spies evs)"
apply (blast dest!: Confidentiality_Kas Confidentiality_Tgs)
done
```

Most general form.

```
lemmas Confidentiality_Tgs_ter = authTicket_authentic
        [THEN Confidentiality_Tgs_bis]

lemmas Confidentiality_Auth_A = authK_authentic
        [THEN Confidentiality_Kas]

lemma Confidentiality_Serv_A:
    "[ Crypt (shrK A) {|Key authK, Agent Tgs, Number Ta, authTicket|}
            ∈ parts (spies evs);
        Crypt authK {|Key servK, Agent B, Number Ts, servTicket|}
            ∈ parts (spies evs);
        ¬ expiredAK Ta evs; ¬ expiredSK Ts evs;
        A ∉ bad;  B ∉ bad; evs ∈ kerbIV ]
    ⟹ Key servK ∉ analz (spies evs)"
apply (drule authK_authentic, assumption, assumption)
apply (blast dest: Confidentiality_Kas Says_Kas_message_form
      servK_authentic_ter Confidentiality_Tgs_bis)
done

lemma Confidentiality_B:
    "[ Crypt (shrK B) {|Agent A, Agent B, Key servK, Number Ts|}
            ∈ parts (spies evs);
        Crypt authK {|Key servK, Agent B, Number Ts, servTicket|}
            ∈ parts (spies evs);
        Crypt (shrK A) {|Key authK, Agent Tgs, Number Ta, authTicket|}
            ∈ parts (spies evs);
        ¬ expiredSK Ts evs; ¬ expiredAK Ta evs;
```

```
           A ∉ bad;   B ∉ bad; B ≠ Tgs; evs ∈ kerbIV ]
        ⟹ Key servK ∉ analz (spies evs)"
apply (frule authK_authentic)
apply (frule_tac [3] Confidentiality_Kas)
apply (frule_tac [6] servTicket_authentic, auto)
apply (blast dest!: Confidentiality_Tgs_bis
        dest: Says_Kas_message_form servK_authentic
        unique_servKeys unique_authKeys)
done
```

The updated protocol makes servkey confidentiality available to B, in the sense of goal availability.

```
lemma u_Confidentiality_B:
     "[ Crypt (shrK B) {|Agent A, Agent B, Key servK, Number Ts|}
          ∈ parts (spies evs);
        ¬ expiredSK Ts evs;
        A ∉ bad;   B ∉ bad;   B ≠ Tgs; evs ∈ kerbIV ]
     ⟹ Key servK ∉ analz (spies evs)"
apply (blast dest: u_servTicket_authentic u_NotexpiredSK_NotexpiredAK
      Confidentiality_Tgs_bis)
done
```

B. Proof Script Fragments for Kerberos V

B.1 Unicity

An authkey is encrypted by one and only one shared key. A servkey is encrypted by one and only one authK.

```
lemma Key_unique_SesKey:
    "⟦ Crypt K  ⦃Key SesKey,  Agent B, T⦄
          ∈ parts (spies evs);
        Crypt K' ⦃Key SesKey,  Agent B', T'⦄
          ∈ parts (spies evs);  Key SesKey ∉ analz (spies evs);
        evs ∈ kerbV ⟧
      ⟹ K=K' & B=B' & T=T'"
apply (erule rev_mp)
apply (erule rev_mp)
apply (erule rev_mp)
apply (erule kerbV.induct, analz_mono_contra)
apply (frule_tac [7] Says_ticket_parts)
apply (frule_tac [5] Says_ticket_parts, simp_all)
```

Fake, K2, K4

```
apply (blast+)
done
```

```
lemma unique_CryptKey:
    "⟦ Crypt (shrK B) ⦃Agent A,  Agent B,  Key SesKey, T⦄
          ∈ parts (spies evs);
        Crypt (shrK B') ⦃Agent A', Agent B', Key SesKey, T'⦄
          ∈ parts (spies evs);  Key SesKey ∉ analz (spies evs);
        evs ∈ kerbIV ⟧
      ⟹ A=A' & B=B' & T=T'"
apply (erule rev_mp)
apply (erule rev_mp)
apply (erule rev_mp)
apply (erule kerbV.induct, analz_mono_contra)
apply (frule_tac [7] Says_ticket_parts)
apply (frule_tac [5] Says_ticket_parts, simp_all)
```

Fake, K2, K4

```
apply (blast+)
done
```

The session key, if secure, uniquely identifies the ticket whether authTicket or servTicket. As a matter of fact, one can read also Tgs in the place of B.

```
lemma unique_authKeys:
    "⟦ Says Kas A
            {Crypt Ka {Key authK, Agent Tgs, Ta}, X} ∈ set evs;
        Says Kas A'
            {Crypt Ka' {Key authK, Agent Tgs, Ta'}, X'} ∈ set evs;
        evs ∈ kerbV ⟧ ⟹ A=A' ∧ Ka=Ka' ∧ Ta=Ta' ∧ X=X'"
apply (erule rev_mp)
apply (erule rev_mp)
apply (erule kerbV.induct)
apply (frule_tac [7] Says_ticket_parts)
apply (frule_tac [5] Says_ticket_parts, simp_all)
apply blast+
done
```

The servkey uniquely identifies the message from Tgs.

```
lemma unique_servKeys:
    "⟦ Says Tgs A
            {Crypt K {Key servK, Agent B, Ts}, X} ∈ set evs;
        Says Tgs A'
            {Crypt K' {Key servK, Agent B', Ts'}, X'} ∈ set evs;
        evs ∈ kerbV ⟧ ⟹ A=A' ∧ B=B' ∧ K=K' ∧ Ts=Ts' ∧ X=X'"
apply (erule rev_mp)
apply (erule rev_mp)
apply (erule kerbV.induct)
apply (frule_tac [7] Says_ticket_parts)
apply (frule_tac [5] Says_ticket_parts, simp_all)
apply blast+
done
```

B.2 Unicity Relying on Timestamps

Novel guarantees, never studied before. Because honest agents always say the right timestamp in authenticators, we can prove unicity guarantees based exactly on timestamps. Classical unicity guarantees are based on nonces. Of course, assuming the agent to be different from the Spy, rather than not in the set bad, would suffice below. Similar guarantees must also hold for Kerberos IV.

Cannot prove a general fact for any message that is sent, but can prove a less general fact concerning only authenticators!

```
lemma honest_never_says_newer_timestamp_in_auth:
    "⟦ (CT evs) ≤ T; Number T ∈ parts {X}; A ∉ bad; evs ∈ kerbV ⟧
    ⟹ Says A B {Y, X} ∉ set evs"
apply (erule rev_mp)
apply (erule kerbV.induct)
```

```
apply (simp_all)
apply force+
done
```

```
lemma honest_never_says_current_timestamp_in_auth:
    "⟦ (CT evs) = T; Number T ∈ parts {X}; A ∉ bad; evs ∈ kerbV ⟧
    ⟹ Says A B {|Y, X|} ∉ set evs"
apply (frule eq_imp_le)
apply (blast dest: honest_never_says_newer_timestamp_in_auth)
done
```

Observe that an honest agent can send the same timestamp on two different traces of the same length, but not on the same trace! A number of theorems with the same formulation and proof follow, all relying on honest_never_says_current_timestamp_in_auth.

```
lemma unique_timestamp_authenticator1:
    "⟦ Says A Kas {|Agent A, Agent Tgs, Number T1|} ∈ set evs;
        Says A Kas' {|Agent A, Agent Tgs', Number T1|} ∈ set evs;
        A ∉bad; evs ∈ kerbV ⟧
    ⟹ Kas=Kas' ∧ Tgs=Tgs'"
apply (erule rev_mp, erule rev_mp)
apply (erule kerbV.induct)
apply (simp_all, blast)
apply auto
apply (simp_all add: honest_never_says_current_timestamp_in_auth)
done
```

```
lemma unique_timestamp_authenticator2:
  "⟦ Says A Tgs {|AT, Crypt AK {|Agent A, Number T2|}, Agent B|}
        ∈ set evs;
      Says A Tgs' {|AT', Crypt AK' {|Agent A, Number T2|}, Agent B'|}
        ∈ set evs;
        A ∉ bad; evs ∈ kerbV ⟧
    ⟹ Tgs=Tgs' ∧ AT=AT' ∧ AK=AK' ∧ B=B'"
apply (erule rev_mp, erule rev_mp)
apply (erule kerbV.induct)
apply (simp_all, blast)
apply auto
apply (simp_all add: honest_never_says_current_timestamp_in_auth)
done
```

```
lemma unique_timestamp_authenticator3:
    "⟦ Says A B {|ST, Crypt SK {|Agent A, Number T|}|} ∈ set evs;
        Says A B' {|ST', Crypt SK' {|Agent A, Number T|}|} ∈ set evs;
        A ∉ bad; evs ∈ kerbV ⟧
    ⟹ B=B' ∧ ST=ST' ∧ SK=SK'"
apply (erule rev_mp, erule rev_mp)
apply (erule kerbV.induct)
apply (simp_all, blast)
apply auto
— The lemma applies as if the second part of the message were an authenticator.
apply (simp_all add: honest_never_says_current_timestamp_in_auth)
```

done

lemma unique_timestamp_authticket:
  "⟦ Says Kas A ⦃X, Crypt (shrK Tgs) ⦃Agent A, Agent Tgs, Key AK, T⦄⦄
        ∈ set evs;
     Says Kas A'
       ⦃X', Crypt (shrK Tgs') ⦃Agent A', Agent Tgs', Key AK', T⦄⦄
        ∈ set evs;
     evs ∈ kerbV ⟧
   ⟹ A=A' ∧ X=X' ∧ Tgs=Tgs' ∧ AK=AK'"
apply (erule rev_mp, erule rev_mp)
apply (erule kerbV.induct)
apply (simp_all)
apply auto
— The lemma applies as if the second part of the message were an authenticator.
apply (simp_all add: honest_never_says_current_timestamp_in_auth)
done

lemma unique_timestamp_servticket:
  "⟦ Says Tgs A ⦃X, Crypt (shrK B) ⦃Agent A, Agent B, Key SK, T⦄⦄
        ∈ set evs;
     Says Tgs A' ⦃X', Crypt (shrK B') ⦃Agent A', Agent B', Key SK', T⦄⦄
        ∈ set evs;
     evs ∈ kerbV ⟧
   ⟹ A=A' ∧ X=X' ∧ B=B' ∧ SK=SK'"
apply (erule rev_mp, erule rev_mp)
apply (erule kerbV.induct)
apply (simp_all)
apply auto
apply (simp_all add: honest_never_says_current_timestamp_in_auth)
done

For the Kas case, we need to inspect the first half of the message, hence we need another lemma, but this only holds for Kas and Tgs.

lemma Kas_never_says_newer_timestamp:
     "⟦ (CT evs) ≤ T; Number T ∈ parts {X}; evs ∈ kerbV ⟧
      ⟹ ∀ A. Says Kas A X ∉ set evs"
apply (erule rev_mp)
apply (erule kerbV.induct)
apply (simp_all)
apply force+
done

lemma Kas_never_says_current_timestamp:
     "⟦ (CT evs) = T; Number T ∈ parts {X}; evs ∈ kerbV ⟧
      ⟹ ∀ A. Says Kas A X ∉ set evs"
apply (frule eq_imp_le)
apply (blast dest: Kas_never_says_newer_timestamp)
done

lemma unique_timestamp_msg2:

```
"⟦ Says Kas A ⦃Crypt (shrK A) ⦃Key AK, Agent Tgs, T⦄, AT⦄
        ∈ set evs;
   Says Kas A' ⦃Crypt (shrK A') ⦃Key AK', Agent Tgs', T⦄, AT'⦄
        ∈ set evs;
     evs ∈ kerbV ⟧
  ⟹ A=A' ∧ AK=AK' ∧ Tgs=Tgs' ∧ AT=AT'"
apply (erule rev_mp, erule rev_mp)
apply (erule kerbV.induct)
apply (simp_all)
apply auto
apply (simp_all add: Kas_never_says_current_timestamp)
done
```

Same argument for the Tgs case.

```
lemma Tgs_never_says_newer_timestamp:
      "⟦ (CT evs) ≤ T; Number T ∈ parts {X}; evs ∈ kerbV ⟧
       ⟹ ∀ A. Says Tgs A X ∉ set evs"
apply (erule rev_mp)
apply (erule kerbV.induct)
apply (simp_all)
apply force+
done
```

```
lemma Tgs_never_says_current_timestamp:
      "⟦ (CT evs) = T; Number T ∈ parts {X}; evs ∈ kerbV ⟧
       ⟹ ∀ A. Says Tgs A X ∉ set evs"
apply (frule eq_imp_le)
apply (blast dest: Tgs_never_says_newer_timestamp)
done
```

```
lemma unique_timestamp_msg4:
  "⟦ Says Tgs A ⦃Crypt (shrK A) ⦃Key SK, Agent B, T⦄, ST⦄
        ∈ set evs;
   Says Tgs A' ⦃Crypt (shrK A') ⦃Key SK', Agent B', T⦄, ST'⦄
        ∈ set evs;
     evs ∈ kerbV ⟧
  ⟹ A=A' ∧ SK=SK' ∧ B=B' ∧ ST=ST'"
apply (erule rev_mp, erule rev_mp)
apply (erule kerbV.induct)
apply (simp_all)
apply auto
apply (simp_all add: Tgs_never_says_current_timestamp)
done
```

B.3 Key Distribution and Non-injective Agreement

Agents' knowledge of session keys. An agent knows a session key if he used it
to issue a cipher. These guarantees can be interpreted both in terms of key
distribution and of non-injective agreement on the session key.

lemma B_Issues_A:
```
    "⟦ Says B A (Crypt servK (Number T3)) ∈ set evs;
        Key servK ∉ analz (spies evs);
        A ∉ bad;  B ∉ bad; B ≠ Tgs; evs ∈ kerbV ⟧
    ⟹ B Issues A with (Crypt servK (Number T3)) on evs"
```
apply (simp (no_asm) add: Issues_def)
apply (rule exI)
apply (rule conjI, assumption)
apply (simp (no_asm))
apply (erule rev_mp)
apply (erule rev_mp)
apply (erule kerbV.induct, analz_mono_contra)
apply (simp_all (no_asm_simp) add: all_conj_distrib)
apply blast

K6 requires numerous lemmas.

apply (simp add: takeWhile_tail)
apply (blast dest: servTicket_authentic parts_spies_takeWhile_mono
        [THEN subsetD] parts_spies_evs_revD2 [THEN subsetD]
        intro: Says_K6)
done

lemma A_authenticates_and_keydist_to_B:
```
    "⟦ Crypt servK (Number T3) ∈ parts (spies evs);
        Crypt authK ⦃Key servK, Agent B, Number Ts⦄
          ∈ parts (spies evs);
        Crypt (shrK A) ⦃Key authK, Agent Tgs, Number Ta⦄
          ∈ parts (spies evs);
        Key authK ∉ analz (spies evs); Key servK ∉ analz (spies evs);
        A ∉ bad;  B ∉ bad; B ≠ Tgs; evs ∈ kerbV ⟧
    ⟹ B Issues A with (Crypt servK (Number T3)) on evs"
```
apply (blast dest!: A_authenticates_B B_Issues_A)
done

lemma A_Issues_B:
```
    "⟦ Says A B ⦃ST, Crypt servK ⦃Agent A, Number T3⦄⦄ ∈ set evs;
        Key servK ∉ analz (spies evs);
        B ≠ Tgs; A ∉ bad;  B ∉ bad;  evs ∈ kerbV ⟧
    ⟹ A Issues B with (Crypt servK ⦃Agent A, Number T3⦄) on evs"
```
apply (simp (no_asm) add: Issues_def)
apply (rule exI)
apply (rule conjI, assumption)
apply (simp (no_asm))
apply (erule rev_mp)
apply (erule rev_mp)
apply (erule kerbV.induct, analz_mono_contra)
apply (frule_tac [7] Says_ticket_parts)
apply (frule_tac [5] Says_ticket_parts)
apply (simp_all (no_asm_simp))

K5

apply auto
apply (simp add: takeWhile_tail)

Case study necessary because the assumption doesn't state the form of servTicket. The guarantee becomes stronger.

```
prefer 2 apply (simp add: takeWhile_tail)
apply (frule K3_imp_K2, assumption, assumption, erule exE, erule exE)
apply (case_tac "Key authK ∈ analz (spies evs5)")
apply (drule Says_imp_knows_Spy [THEN analz.Inj, THEN analz.Fst,
THEN analz_Decrypt', THEN analz.Fst], assumption, assumption, simp)
apply (frule K3_imp_K2, assumption, assumption, erule exE, erule exE)
apply (drule Says_imp_knows_Spy [THEN parts.Inj, THEN parts.Fst])
apply (frule servK_authentic_ter, blast, assumption+)
apply (drule parts_spies_takeWhile_mono [THEN subsetD])
apply (drule parts_spies_evs_revD2 [THEN subsetD])
```

Says_K5 closes the proof in version IV because it is clear which servTicket an authenticator appears with in msg 5. In version V an authenticator can appear with any item that the Spy could replace the servTicket with.

```
apply (frule Says_K5, blast, assumption, assumption, assumption,
       assumption, erule exE)
```

We need to state that an honest agent wouldn't send the wrong timestamp within an authenticator, wathever it is paired with.

```
apply (simp add: honest_never_says_current_timestamp_in_auth)
done

lemma B_authenticates_and_keydist_to_A:
    "⟦ Crypt servK ⦃Agent A, Number T3⦄ ∈ parts (spies evs);
        Crypt (shrK B) ⦃Agent A, Agent B, Key servK, Number Ts⦄
          ∈ parts (spies evs);
        Key servK ∉ analz (spies evs);
        B ≠ Tgs; A ∉ bad;  B ∉ bad;  evs ∈ kerbV ⟧
   ⟹ A Issues B with (Crypt servK ⦃Agent A, Number T3⦄) on evs"
apply (blast dest: B_authenticates_A A_Issues_B)
done
```

C. Proof Script Fragments for Shoup-Rubin

C.1 Function "initState"

consts
```
  initState :: "agent => msg set"
```
primrec

```
  initState_Server:  "initState Server =
     (Key'(range shrK ∪ range crdK ∪ range PIN ∪ range pairK)) ∪
     (Nonce'(range Pairkey))"


  initState_Friend: "initState (Friend i) = {Key (PIN (Friend i))}"


  initState_Spy: "initState Spy  =
                  (Key'((PIN'bad) ∪ (PIN '{A. Card A ∈ cloned}) ∪
                                   (shrK'{A. Card A ∈ cloned}) ∪
                   (crdK'cloned) ∪
                   (pairK'{(X,A). Card A ∈ cloned}))) ∪
     ∪ (Nonce'(Pairkey'{(A,B). Card A ∈ cloned & Card B ∈ cloned}))"
```

lemma shrK_in_initState [iff]: "Key (shrK A) ∈ initState Server"
apply (induct_tac "A")
apply auto
done

lemma shrK_in_used [iff]: "Key (shrK A) ∈ used evs"
apply (rule initState_into_used)
apply blast
done

lemma crdK_in_initState [iff]: "Key (crdK A) ∈ initState Server"
apply (induct_tac "A")
apply auto
done

lemma crdK_in_used [iff]: "Key (crdK A) ∈ used evs"
apply (rule initState_into_used)
apply blast
done

```
lemma PIN_in_initState [iff]: "Key (PIN A) ∈ initState A"
apply (induct_tac "A")
apply auto
done

lemma PIN_in_used [iff]: "Key (PIN A) ∈ used evs"
apply (rule initState_into_used)
apply blast
done

lemma pairK_in_initState [iff]: "Key (pairK X) ∈ initState Server"
apply (induct_tac "X")
apply auto
done

lemma pairK_in_used [iff]: "Key (pairK X) ∈ used evs"
apply (rule initState_into_used)
apply blast
done
```

C.2 Function "knows"

```
consts
  knows    :: "agent => event list => msg set"


primrec
  knows_Nil:    "knows A [] = initState A"
  knows_Cons:   "knows A (ev # evs) =
    (case ev of
       Says A' B X =>
           if (A=A' | A=Spy) then insert X (knows A evs)
           else knows A evs
     | Notes A' X  =>
           if (A=A' | (A=Spy & A'∈bad)) then insert X (knows A evs)
                                        else knows A evs
     | Gets A' X   =>
           if (A=A' & A ≠ Spy) then insert X (knows A evs)
                               else knows A evs
     | Inputs A' C X =>
         if secureM then
           if A=A' then insert X (knows A evs) else knows A evs
         else
           if (A=A' | A=Spy) then insert X (knows A evs)
           else knows A evs
     | C_Gets C X   => knows A evs
     | Outpts C A' X =>
         if secureM then
           if A=A' then insert X (knows A evs) else knows A evs
         else
           if A=Spy then insert X (knows A evs) else knows A evs
     | A_Gets A' X   =>
```

```
                if (A=A' & A ≠ Spy) then insert X (knows A evs)
                              else knows A evs)"
```

lemma knows_Spy_Says [simp]:
    "knows Spy (Says A B X # evs) = insert X (knows Spy evs)"
by simp

Letting the Spy see compromised agents' notes avoids redundant case-splits on whether A is the Spy and whether she is compromised.

lemma knows_Spy_Notes [simp]:
    "knows Spy (Notes A X # evs) =
        (if A∈ bad then insert X (knows Spy evs) else knows Spy evs)"
by simp

lemma knows_Spy_Gets [simp]:
        "knows Spy (Gets A X # evs) = knows Spy evs"
by simp

lemma knows_Spy_Inputs_secureM [simp]:
    "secureM ⟹ knows Spy (Inputs A C X # evs) =
      (if A=Spy then insert X (knows Spy evs) else knows Spy evs)"
by simp

lemma knows_Spy_Inputs_insecureM [simp]:
    "insecureM ⟹
        knows Spy (Inputs A C X # evs) = insert X (knows Spy evs)"
by simp

lemma knows_Spy_C_Gets [simp]:
        "knows Spy (C_Gets C X # evs) = knows Spy evs"
by simp

lemma knows_Spy_Outpts_secureM [simp]:
    "secureM ⟹ knows Spy (Outpts C A X # evs) =
        (if A=Spy then insert X (knows Spy evs) else knows Spy evs)"
by simp

lemma knows_Spy_Outpts_insecureM [simp]:
    "insecureM ⟹
        knows Spy (Outpts C A X # evs) = insert X (knows Spy evs)"
by simp

lemma knows_Spy_A_Gets [simp]:
        "knows Spy (A_Gets A X # evs) = knows Spy evs"
by simp

C.3 Authentication

lemma Outpts_A_Card_form_10:
    "⟦ Outpts (Card A) A {Key K, Certificate} ∈ set evs; evs ∈ sr ⟧

```
        ⟹ ∃ B Nb.
            K = sesK(Nb,pairK(A,B)) ∧
            Certificate = (Crypt (pairK(A,B)) (Nonce Nb))"
apply (erule rev_mp, erule sr.induct)
apply (simp_all (no_asm_simp))
done

lemma Na_Nb_certificate_authentic:
  "⟦ Crypt (pairK(A,B)) ⦃Nonce Na, Nonce Nb⦄ ∈ parts (knows Spy evs);
        ¬illegalUse(Card B);
          evs ∈ sr ⟧
      ⟹ Outpts (Card B) B ⦃Nonce Nb, Key (sesK(Nb,pairK(A,B))),
                Crypt (pairK(A,B)) ⦃Nonce Na, Nonce Nb⦄,
                Crypt (pairK(A,B)) (Nonce Nb)⦄ ∈ set evs"
apply (erule rev_mp, erule sr.induct)
apply parts_prepare
apply simp_all
apply spy_analz
apply clarify
done

lemma Nb_certificate_authentic:
      "⟦ Crypt (pairK(A,B)) (Nonce Nb) ∈ parts (knows Spy evs);
        B ≠ Spy; ¬illegalUse(Card A); ¬illegalUse(Card B);
          evs ∈ sr ⟧
      ⟹ Outpts (Card A) A ⦃Key (sesK(Nb,pairK(A,B))),
                          Crypt (pairK(A,B)) (Nonce Nb)⦄ ∈ set evs"
apply (erule rev_mp, erule sr.induct)
apply parts_prepare
apply (case_tac [17] "Aa = Spy")
apply simp_all
apply spy_analz
apply clarify+
done

lemma A_authenticates_B:
  "⟦ Outpts (Card A) A ⦃Key K, Crypt (pairK(A,B)) (Nonce Nb)⦄
        ∈ set evs;
      ¬illegalUse(Card B);
      evs ∈ sr ⟧
      ⟹ ∃ Na.
            Outpts (Card B) B ⦃Nonce Nb, Key K,
                Crypt (pairK(A,B)) ⦃Nonce Na, Nonce Nb⦄,
                Crypt (pairK(A,B)) (Nonce Nb)⦄ ∈ set evs"
apply (blast dest: Na_Nb_certificate_authentic Outpts_A_Card_form_10
Outpts_A_Card_imp_pairK_parts)
done

lemma A_authenticates_B_Gets:
      "⟦ Gets A ⦃Nonce Nb, Crypt (pairK(A,B)) ⦃Nonce Na, Nonce Nb⦄⦄
          ∈ set evs;
        ¬illegalUse(Card B);
          evs ∈ sr ⟧
```

```
    ⟹ Outpts (Card B) B ⦃Nonce Nb, Key (sesK(Nb, pairK (A, B))),
                        Crypt (pairK(A,B)) ⦃Nonce Na, Nonce Nb⦄,
                        Crypt (pairK(A,B)) (Nonce Nb)⦄ ∈ set evs"
apply (blast dest: Gets_imp_knows_Spy [THEN parts.Inj, THEN parts.Snd,
      THEN Na_Nb_certificate_authentic])
done

lemma B_authenticates_A:
    "⟦ Gets B (Crypt (pairK(A,B)) (Nonce Nb)) ∈ set evs;
       B ≠ Spy; ¬illegalUse(Card A); ¬illegalUse(Card B);
       evs ∈ sr ⟧
     ⟹ Outpts (Card A) A
     ⦃Key (sesK(Nb,pairK(A,B))), Crypt (pairK(A,B)) (Nonce Nb)⦄
     ∈ set evs"
apply (erule rev_mp)
apply (erule sr.induct)
apply (simp_all (no_asm_simp))
apply (blast dest: Says_imp_knows_Spy [THEN parts.Inj]
      Nb_certificate_authentic)
done
```

D. Proof Script Fragments for Zhou-Gollmann

D.1 Validity of Main Evidence

Below, we prove that if NRO exists, then A definitely sent it, provided A is not broken.

Strong conclusion for a good agent.

```
lemma NRO_validity_good:
    "⟦NRO = Crypt (priK A) ⦃Number f_nro, Agent B, Nonce L, C⦄;
      NRO ∈ parts (spies evs);
      A ∉ bad;  evs ∈ zg ⟧
      ⟹ Says A B ⦃Number f_nro, Agent B, Nonce L, C, NRO⦄ ∈ set evs"
apply clarify
apply (erule rev_mp)
apply (erule zg.induct)
apply (frule_tac [5] ZG2_msg_in_parts_spies, auto)
done

lemma NRO_sender:
    "⟦Says A' B ⦃n, b, l, C, Crypt (priK A) X⦄ ∈ set evs; evs ∈ zg⟧
      ⟹ A' ∈ {A,Spy}"
apply (erule rev_mp)
apply (erule zg.induct, simp_all)
done
```

Holds also for A = Spy!

```
theorem NRO_validity:
    "⟦Gets B ⦃Number f_nro, Agent B, Nonce L, C, NRO⦄ ∈ set evs;
      NRO = Crypt (priK A) ⦃Number f_nro, Agent B, Nonce L, C⦄;
      A ∉ broken;  evs ∈ zg ⟧
      ⟹ Says A B ⦃Number f_nro, Agent B, Nonce L, C, NRO⦄ ∈ set evs"
apply (drule Gets_imp_Says, assumption)
apply clarify
apply (frule NRO_sender, auto)
```

We are left with the case where the sender is Spy and not equal to A, because A is uncompromised. Thus, Theorem NRO_validity_good applies.

```
apply (blast dest: NRO_validity_good [OF refl])
done
```

Below, we prove that if NRR exists, then B definitely sent it, provided B is not broken.

Strong conclusion for a good agent.

lemma NRR_validity_good:
    "⟦NRR = Crypt (priK B) ⦃Number f_nrr, Agent A, Nonce L, C⦄;
        NRR ∈ parts (spies evs);
        B ∉ bad; evs ∈ zg ⟧
    ⟹ Says B A ⦃Number f_nrr, Agent A, Nonce L, NRR⦄ ∈ set evs"
apply clarify
apply (erule rev_mp)
apply (erule zg.induct)
apply (frule_tac [5] ZG2_msg_in_parts_spies, auto)
done

lemma NRR_sender:
    "⟦Says B' A ⦃n, a, l, Crypt (priK B) X⦄ ∈ set evs; evs ∈ zg⟧
    ⟹ B' ∈ {B,Spy}"
apply (erule rev_mp)
apply (erule zg.induct, simp_all)
done

Holds also for B = Spy!

theorem NRR_validity:
    "⟦Says B' A ⦃Number f_nrr, Agent A, Nonce L, NRR⦄ ∈ set evs;
        NRR = Crypt (priK B) ⦃Number f_nrr, Agent A, Nonce L, C⦄;
        B ∉ broken; evs ∈ zg⟧
    ⟹ Says B A ⦃Number f_nrr, Agent A, Nonce L, NRR⦄ ∈ set evs"
apply clarify
apply (frule NRR_sender, auto)

We are left with the case where B' = Spy and B' is not B, namely B is uncompromised, when we can apply NRR_validity_good.

apply (blast dest: NRR_validity_good [OF refl])
done

D.2 Validity of Subsidiary Evidence

Below, we prove that if sub_K exists, then A definitely sent it, provided A is not broken.

Strong conclusion for a good agent.

lemma sub_K_validity_good:
    "⟦sub_K = Crypt (priK A) ⦃Number f_sub, Agent B, Nonce L, Key K⦄;
        sub_K ∈ parts (spies evs);
        A ∉ bad; evs ∈ zg ⟧
    ⟹ Says A TTP ⦃Number f_sub, Agent B, Nonce L, Key K, sub_K⦄
        ∈ set evs"
apply clarify

```
apply (erule rev_mp)
apply (erule zg.induct)
apply (frule_tac [5] ZG2_msg_in_parts_spies, simp_all)
```

Fake

```
apply (blast dest!: Fake_parts_sing_imp_Un)
done
```

```
lemma sub_K_sender:
    "⟦Says A' TTP ⦃n, b, l, k, Crypt (priK A) X⦄ ∈ set evs;  evs ∈ zg⟧
       ⟹ A' ∈ {A,Spy}"
apply (erule rev_mp)
apply (erule zg.induct, simp_all)
done
```

Holds also for A = Spy!

```
theorem sub_K_validity:
 "⟦Gets TTP ⦃Number f_sub, Agent B, Nonce L, Key K, sub_K⦄ ∈ set evs;
    sub_K = Crypt (priK A) ⦃Number f_sub, Agent B, Nonce L, Key K⦄;
    A ∉ broken;  evs ∈ zg ⟧
 ⟹ Says A TTP ⦃Number f_sub, Agent B, Nonce L, Key K, sub_K⦄
       ∈ set evs"
apply (drule Gets_imp_Says, assumption)
apply clarify
apply (frule sub_K_sender, auto)
```

We are left with the case where the sender is Spy and not equal to A, because A is uncompromised. Thus, Theorem sub_K_validity_good applies.

```
apply (blast dest: sub_K_validity_good [OF refl])
done
```

Below, we prove that if con_K exists, then TTP has it, and therefore A and B can get it too. Moreover, we know that A sent sub_K.

```
lemma con_K_validity:
    "⟦con_K ∈ used evs;
        con_K = Crypt (priK TTP)
                    ⦃Number f_con, Agent A, Agent B, Nonce L, Key K⦄;
        evs ∈ zg ⟧
 ⟹ Notes TTP ⦃Number f_con, Agent A, Agent B, Nonce L, Key K, con_K⦄
         ∈ set evs"
apply clarify
apply (erule rev_mp)
apply (erule zg.induct)
apply (frule_tac [5] ZG2_msg_in_parts_spies, simp_all)
```

Fake

```
apply (blast dest!: Fake_parts_sing_imp_Un)
```

ZG2

```
apply (blast dest: parts_cut)
done
```

If TTP holds con_K, then A sent sub_K. We assume that A is not broken. Importantly, nothing needs to be assumed about the form of con_K!

```
lemma Notes_TTP_imp_Says_A:
 "⟦Notes TTP ⦃Number f_con, Agent A, Agent B, Nonce L, Key K, con_K⦄
      ∈ set evs;
    sub_K = Crypt (priK A) ⦃Number f_sub, Agent B, Nonce L, Key K⦄;
    A ∉ broken; evs ∈ zg⟧
  ⟹ Says A TTP ⦃Number f_sub, Agent B, Nonce L, Key K, sub_K⦄
      ∈ set evs"
apply clarify
apply (erule rev_mp)
apply (erule zg.induct)
apply (frule_tac [5] ZG2_msg_in_parts_spies, simp_all)
```

ZG4

```
apply clarify
apply (rule sub_K_validity, auto)
done
```

If con_K exists, then A sent sub_K. We again assume that A is not broken.

```
theorem B_sub_K_validity:
   "⟦con_K ∈ used evs;
     con_K = Crypt (priK TTP) ⦃Number f_con, Agent A, Agent B,
                                Nonce L, Key K⦄;
     sub_K = Crypt (priK A) ⦃Number f_sub, Agent B, Nonce L, Key K⦄;
     A ∉ broken; evs ∈ zg⟧
  ⟹ Says A TTP ⦃Number f_sub, Agent B, Nonce L, Key K, sub_K⦄
       ∈ set evs"
by (blast dest: con_K_validity Notes_TTP_imp_Says_A)
```

D.3 Fairness

Cannot prove that, if B has NRO, then A has her NRR. It would appear that B has a small advantage, though it is useless to win disputes: B needs to present con_K as well.

```
lemma A_unicity:
   "⟦NRO = Crypt (priK A) ⦃Number f_nro, Agent B, Nonce L, Crypt K M⦄;
     NRO ∈ parts (spies evs);
       Says A B ⦃Number f_nro, Agent B, Nonce L, Crypt K M', NRO'⦄
         ∈ set evs;
       A ∉ bad; evs ∈ zg ⟧
    ⟹ M'=M"
apply clarify
apply (erule rev_mp)
apply (erule rev_mp)
apply (erule zg.induct)
apply (frule_tac [5] ZG2_msg_in_parts_spies, auto)
```

ZG1: freshness

```
apply (blast dest: parts.Body)
done
```

Fairness lemma: if sub_K exists, then A holds NRR. Relies on unicity of labels.

```
lemma sub_K_implies_NRR:
   "⟦ NRO = Crypt (priK A) {|Number f_nro, Agent B, Nonce L, Crypt K M|};
      NRR = Crypt (priK B) {|Number f_nrr, Agent A, Nonce L, Crypt K M|};
      sub_K ∈ parts (spies evs);
      NRO ∈ parts (spies evs);
      sub_K = Crypt (priK A) {|Number f_sub, Agent B, Nonce L, Key K|};
      A ∉ bad;   evs ∈ zg ⟧
⟹ Gets A {|Number f_nrr, Agent A, Nonce L, NRR|} ∈ set evs"
apply clarify
apply (erule rev_mp)
apply (erule rev_mp)
apply (erule zg.induct)
apply (frule_tac [5] ZG2_msg_in_parts_spies, simp_all)
```

Fake

```
apply blast
```

ZG1: freshness

```
apply (blast dest: parts.Body)
```

ZG3

```
apply (blast dest: A_unicity [OF refl])
done
```

```
lemma Crypt_used_imp_L_used:
    "⟦ Crypt (priK TTP) {|F, A, B, L, K|} ∈ used evs; evs ∈ zg ⟧
       ⟹ L ∈ used evs"
apply (erule rev_mp)
apply (erule zg.induct, auto)
```

Fake

```
apply (blast dest!: Fake_parts_sing_imp_Un)
```

ZG2: freshness

```
apply (blast dest: parts.Body)
done
```

Fairness for A: if con_K and NRO exist, then A holds NRR. A must be uncompromised, but there is no assumption about B.

```
theorem A_fairness_NRO:
   "⟦con_K ∈ used evs;
     NRO ∈ parts (spies evs);
     con_K = Crypt (priK TTP)
                   {|Number f_con, Agent A, Agent B, Nonce L, Key K|};
     NRO = Crypt (priK A) {|Number f_nro, Agent B, Nonce L, Crypt K M|};
     NRR = Crypt (priK B) {|Number f_nrr, Agent A, Nonce L, Crypt K M|};
```

```
      A ∉ bad;  evs ∈ zg ]
      ⟹ Gets A {Number f_nrr, Agent A, Nonce L, NRR} ∈ set evs"
apply clarify
apply (erule rev_mp)
apply (erule rev_mp)
apply (erule zg.induct)
apply (frule_tac [5] ZG2_msg_in_parts_spies, simp_all)
```

Fake

```
  apply (simp add: parts_insert_knows_A)
  apply (blast dest: Fake_parts_sing_imp_Un)
```

ZG1

```
  apply (blast dest: Crypt_used_imp_L_used)
```

ZG2

```
apply (blast dest: parts_cut)
```

ZG4

```
apply (blast intro: sub_K_implies_NRR [OF refl]
            dest: Gets_imp_knows_Spy [THEN parts.Inj])
done
```

Fairness for B: NRR exists at all, then B holds NRO. B must be uncompromised, but there is no assumption about A.

```
theorem B_fairness_NRR:
    "[NRR ∈ used evs;
      NRR = Crypt (priK B) {Number f_nrr, Agent A, Nonce L, C};
      NRO = Crypt (priK A) {Number f_nro, Agent B, Nonce L, C};
      B ∉ bad; evs ∈ zg ]
    ⟹ Gets B {Number f_nro, Agent B, Nonce L, C, NRO} ∈ set evs"
apply clarify
apply (erule rev_mp)
apply (erule zg.induct)
apply (frule_tac [5] ZG2_msg_in_parts_spies, simp_all)
```

Fake

```
apply (blast dest!: Fake_parts_sing_imp_Un)
```

ZG2

```
apply (blast dest: parts_cut)
done
```

If con_K exists at all, then B can get it, by con_K_validity. We cannot conclude that also NRO is available to B, because if A were unfair, A could build message 3 without building message 1, which contains NRO.

Bibliography

1. M. Abadi. Secrecy by typing in security protocols. *Journal of the ACM*, 46(5):749–786, 1999.
2. M. Abadi and B. Blanchet. Computer-assisted verification of a protocol for certified email. In R. Cousot, editor, *Proc. of the 10th International Symposium on Static Analysis (SAS'03)*, LNCS 2694, pages 316–335. Springer-Verlag, 2003.
3. M. Abadi, M. Burrows, C. Kaufman, and B. Lampson. Authentication and delegation with smart-cards. Research Report 125, Digital - Systems Research Center, 1994.
4. M. Abadi, N. Glew, B. Horne, and B. Pinkas. Certified email with a light on-line trusted third party: Design and implementation. In *Proc. of the 11th International Conference on World Wide Web (WWW'02)*, pages 387–395. ACM Press, 2002.
5. M. Abadi and A. Gordon. Reasoning about cryptographic protocols in the spi calculus. In A. W. Mazurkiewicz and J. Winkowski, editors, *Proc. of the 8th International Conference on Concurrency Theory (CONCUR'97)*, LNCS 1243, pages 59–73. Springer-Verlag, 1997.
6. M. Abadi and A. Gordon. A calculus for cryptographic protocols: the spi calculus. *Information and Computation*, 148(1):1–70, 1999.
7. M. Abadi and R. M. Needham. Prudent engineering practice for cryptographic protocols. *IEEE Transactions on Software Engineering*, 22(1):6–15, 1996.
8. M. Abadi and M. Tuttle. A semantics for a logic of authentication. In *Proc. of the 10th ACM Symposium on Principles of Distributed Computing (PODC'91)*, pages 201–216. ACM Press, 1991.
9. M. Abdalla, P. A. Fouque, and D. Pointcheval. Password-based authenticated key exchange in the three-party setting. *IEE Proceedings Information Security*, 153(1):27–39, 2006.
10. J. Alves-Foss. Multiprotocol attacks and the public key infrastructure. In *Proc. of the 21st National Information Systems Security Conference*, pages 566–576, 1998.
11. N. Amla, X. Du, A. Kuehlmann, R. P. Kurshan, and K. L. McMillan. An analysis of SAT-based model checking techniques in an industrial environment. In D. Borrione and W. J. Paul, editors, *Proc. of the 13th Conference on Correct Hardware Design and Verification Methods (CHARME'05)*, LNCS 3725, pages 254–268. Springer-Verlag, 2005.
12. R. Anderson. Why cryptosystems fail. In *Proc. of the 1st ACM Conference on Communications and Computer Security (CCS'93)*, pages 217–227. ACM Press, 1993.
13. R. Anderson and R. M. Needham. Programming Satan's computer. In J. Van Leeuwen, editor, *Computer Science Today: Recent Trends and Developments*, LNCS 1000, pages 426–441. Springer-Verlag, 1995.

14. R. Anderson and R. M. Needham. Robustness principles for public key protocols. In D. Coppersmith, editor, *Proc. of Advances in Cryptography (CRYPTO'95)*, LNCS 963, pages 236–247. Springer-Verlag, 1995.

15. R. J. Anderson and M. J. Kuhn. Low cost attacks on tamper resistant devices. In B. Christianson, B. Crispo, T. M. A. Lomas, and M. Roe, editors, *Proc. of the 5th Security Protocols Workshop (SPW'97)*, LNCS 1361, pages 125–136. Springer-Verlag, 1998.

16. A. Armando and L. Compagna. An optimized intruder model for SAT-based model-checking of security protocols. In *Proc. of the Workshop on Automated Reasoning for Security Protocol Analysis (ARSPA'04)*, ENTCS 125, pages 91–108. Elsevier Science, 2005.

17. N. Asokan, V. Shoup, and M. Waidner. Asynchronous protocols for optimistic fair exchange. In *Proc. of the 17th IEEE Symposium on Security and Privacy (SSP'98)*, pages 86–99. IEEE Press, 1998.

18. G. Ateniese, M. Steiner, and G. Tsudik. Authenticated group key agreement and friends. In *Proc. of the 5th ACM Conference on Computer and Communication Security (CCS'98)*, pages 17–26. ACM Press, 1998.

19. M. Barjaktarovic, C. Shiu-Kai, J. Faust, C. Hosmer, D. Rosenthal, M. Stillman, G. Hird, and D. Zhou. Analysis and implementation of secure electronic mail protocols. In H. Orman and C. Meadows, editors, *Proc. of the Workshop on Design and Formal Verification of Security Protocols (DIMACS'97)*, 1997.

20. D. Basin, S. Mödersheim, and L. Viganò. OFMC: A symbolic model-checker for security protocols. *International Journal of Information Security*, 4(3):181–208, 2005.

21. D. A. Basin, S. Mödersheim, and L. Viganò. An on-the-fly model-checker for security protocol analysis. In E. Snekkenes and D. Gollmann, editors, *Proc. of the 8th European Symposium on Research in Computer Security (ES-ORICS'03)*, LNCS 2808, pages 253–270. Springer-Verlag, 2003.

22. G. Bella. Message reception in the inductive approach. Technical Report 460, Computer Laboratory, University of Cambridge, 1999.

23. G. Bella. Inductive verification of cryptographic protocols. Ph.D. Thesis. Technical Report 493, Cambridge University Computer Laboratory, 2000.

24. G. Bella. Modelling agents' knowledge inductively. In B. Christianson, B. Crispo, J. A. Malcolm, and R. Michael, editors, *Proc. of the 7th Security Protocols Workshop (SPW'99)*, LNCS 1796, pages 85–94. Springer-Verlag, 2000.

25. G. Bella. Lack of explicitness strikes back. In *Proc. of the 8th Security Protocols Workshop (SPW'00)*, LNCS 2133, pages 87–99. Springer-Verlag, 2001.

26. G. Bella. Mechanising a protocol for smartcards. In I. Attali and T. Jensen, editors, *Proc. of the 1st International Conference on Research in Smartcards (e-Smart'01)*, LNCS 2140, pages 19–33. Springer-Verlag, 2001.

27. G. Bella. Availability of protocol goals. In B. Panda, editor, *Proc. of the 18th ACM Symposium on Applied Computing (ACM SAC'03)*, pages 312–317. ACM Press, 2003.

28. G. Bella. Inductive verification of smartcard protocols. *Journal of Computer Security*, 11(1):87–132, 2003.

29. G. Bella and S. Bistarelli. Soft constraints for security protocol analysis: Confidentiality. In I. V. Ramakrishnan, editor, *Proc. of the 3rd International Symposium on Practical Aspects of Declarative Languages (PADL'01)*, LNCS 1990, pages 108–122. Springer-Verlag, 2001.

30. G. Bella and S. Bistarelli. Soft constraint programming to analysing security protocols. *Theory and Practice of Logic Programming*, 4(5):1–28, 2004.

31. G. Bella, S. Bistarelli, and S. N. Foley. Soft constraints for security. In M. ter Beek and F. Gadducci, editors, *Proc. of the 1st International Workshop on Views On Designing Complex Architectures (VODCA'04)*, ENTCS 142, pages 11–29. Elsevier Science, 2004.

32. G. Bella, S. Bistarelli, and F. Massacci. A protocol's life after attacks. In *Proc. of the 11th Security Protocols Workshop (SPW'03)*, LNCS 3364, pages 3–18. Springer-Verlag, 2005.

33. G. Bella, F. Blanqui, and L. C. Paulson. *Local Repository of Protocol Proofs*, As from Isabelle 2006. Isabelle's local subdirectory /src/HOL/Auth.

34. G. Bella, F. Blanqui, and L. C. Paulson. *On-line Repository of Protocol Proofs*, As from Isabelle 2006.
    http://isabelle.in.tum.de/library/HOL/Auth/index.html.

35. G. Bella, C. Longo, and L. C. Paulson. Verifying second-level security protocols. In D. Basin and B. Wolff, editors, *Proc. of the 16th International Conference on Theorem Proving in Higher Order Logics (TPHOLs'03)*, LNCS 2758, pages 352–366. Springer-Verlag, 2003.

36. G. Bella, F. Massacci, and L. C. Paulson. Verifying the SET registration protocols. *IEEE Journal of Selected Areas in Communications*, 21(1):77–87, 2003.

37. G. Bella, F. Massacci, and L. C. Paulson. An overview of the verification of SET. *International Journal of Information Security*, 4(1-2):17–28, 2005.

38. G. Bella, F. Massacci, and L. C. Paulson. Verifying the SET purchase protocols. *Journal of Automated Reasoning*, 36(1-2):5–37, 2006.

39. G. Bella and L. C. Paulson. Are timestamps worth the effort? A formal treatment. Technical Report 447, Cambridge University Computer Laboratory, 1998.

40. G. Bella and L. C. Paulson. Kerberos Version IV: Inductive analysis of the secrecy goals. In J.-J. Quisquater, Y. Deswarte, C. Meadows, and D. Gollmann, editors, *Proc. of the 5th European Symposium on Research in Computer Security (ESORICS'98)*, LNCS 1485, pages 361–375. Springer-Verlag, 1998.

41. G. Bella and L. C. Paulson. Mechanising BAN Kerberos by the Inductive Method. In A. J. Hu and M. Y. Vardi, editors, *Proc. of the International Conference on Computer-Aided Verification (CAV'98)*, LNCS 1427, pages 416–427. Springer-Verlag, 1998.

42. G. Bella and L. C. Paulson. Mechanical proofs about a non-repudiaton protocol. In R. J. Boulton and P. B. Jackson, editors, *Proc. of the 14th International Conference on Theorem Proving in Higher Order Logics (TPHOLs'01)*, LNCS 2152, pages 91–104. Springer-Verlag, 2001.

43. G. Bella and L. C. Paulson. Analyzing delegation properties. In B. Christianson, B. Crispo, W. S. Harbison, and M. Roe, editors, *Proc. of the 10th Security Protocols Workshop (SPW'02)*, LNCS 2845, pages 120–127. Springer-Verlag, 2004.

44. G. Bella and L. C. Paulson. Accountability protocols: Formalized and verified. *ACM Transactions on Information and System Security*, 9(2):1–24, 2006.

45. G. Bella and E. Riccobene. Formal analysis of the Kerberos authentication system. *Journal of Universal Computer Science*, 3(12):1337–1381, 1997.

46. G. Bella and E. Riccobene. A realistic environment for crypto-protocol analyses by ASMs. In U. Glässer, editor, *Proc. of the 5th International Workshop on Abstract State Machines (Informatik'98)*, pages 127–138, 1998.

47. M. Bellare and P. Rogaway. Entity authentication and key distribution. In D. R. Stinson, editor, *Proc. of Advances in Cryptography (CRYPTO'93)*, LNCS 773, pages 232–249. Springer-Verlag, 1993.

48. M. Bellare and P. Rogaway. Provably secure session key distribution — the three party case. In *Proc. of the 27th ACM SIGACT Symposium on Theory of Computing (STOC'95)*, pages 57–66. ACM Press, 1995.

49. S. Bistarelli. *Semirings for Soft Constraint Solving and Programming*. LNCS 2962. Springer-Verlag, 2004.

50. S. Bistarelli and S. N. Foley. A constraint framework for the qualitative analysis of dependability goals: Integrity. In S. Anderson, M. Felici, and B. Littlewood, editors, *Proc. of the 22nd International Conference on Computer Safety, Reliability and Security (SAFECOMP'03)*, LNCS 2788, pages 130–143. Springer-Verlag, 2003.

51. B. Blanchet. An efficient cryptographic protocol verifier based on Prolog rules. In *Proc. of the 14th IEEE Computer Security Foundations Workshop (CSFW'01)*, pages 82–96. IEEE Press, 1998.

52. A. Bleeker and L. Meertens. A semantics for BAN logic. In H. Orman and C. Meadows, editors, *Proc. of the Workshop on Design and Formal Verification of Security Protocols (DIMACS'97)*, 1997.

53. E. Börger. Annotated bibliography on Evolving Algebras. In E. Börger, editor, *Specification and Validation Methods*, pages 37–52. Oxford University Press, 1995.

54. E. Börger and L. Mearelli. Integrating ASMs into the software development life cycle. *Journal of Universal Computer Science*, 3(5):603–665, 1997.

55. C. Boyd and A. Mathuria. *Protocols for Authentication and Key Establishment*. Information Security and Cryptography Series. Springer-Verlag, 2003.

56. S. Brackin. A HOL extension of GNY for automatically cryptographic protocols. In *Proc. of the 9th IEEE Computer Security Foundations Workshop (CSFW'96)*, pages 62–76. IEEE Press, 1996.

57. S. H. Brackin. Automatic formal analyses of two large commercial protocols. In H. Orman and C. Meadows, editors, *Proc. of the Workshop on Design and Formal Verification of Security Protocols (DIMACS'97)*, 1997.

58. M. Burrows, M. Abadi, and R. M. Needham. A logic of authentication. *Proc. of the Royal Society of London*, 426:233–271, 1989.

59. C. Caleiro, L. Viganò, and D. Basin. Relating strand spaces and distributed temporal logic for security protocol analysis. *Logic Journal of the Interest Group in Pure and Applied Logics*, 13(6):637–663, 2005.

60. I. Cervesato, C. Meadows, and D. Pavlovic. An encapsulated authentication logic for reasoning about key distribution protocols. In *Proc. of the 18th IEEE Computer Security Foundations Workshop (CSFW'05)*, pages 48–61. IEEE Press, 2005.

61. J. Clark and J. Jacob. A survey of authentication protocol literature: Version 1.0. Technical report, University of York, Department of Computer Science, November 1997. http://www-users.cs.york.ac.uk/~jac/.

62. E. M. Clarke, S. Jha, and W. Marrero. Using state space exploration and a natural deduction style message derivation engine to verify security protocols. In D. Gries and W. P. De Roever, editors, *Proc. of the IFIP Working Conference on Programming Concepts and Methods (PROCOMET'98)*, pages 87–106. Chapman & Hall, 1998.

63. E. M. Clarke, S. Jha, and W. Marrero. Verifying security protocols with Brutus. *ACM Transactions on Software Engineering and Methodology*, 9(4):443–487, 2000.

64. E. Cohen. First-order verification of cryptographic protocols. *Journal of Computer Security*, 11(2):186–216, 2003.

65. R. Corin and S. Etalle. An improved constraint-based system for the verification of security protocols. In M. V. Hermenegildo and G. Puebla, editors,

Proc. of the 9th International Symposium on Static Analysis (SAS'02), LNCS 2477, pages 326–341. Springer-Verlag, 2002.

66. C. J. F. Cremers. Feasibility of multi-protocol attacks. In *Proc. of the First International Conference on Availability, Reliability and Security (ARES'06)*, pages 287–294. IEEE Press, 2006.

67. B. Crispo. Delegation protocols for electronic commerce. In *Proc. of the 6th Symposium on Computers and Communications (ISCC'01)*. IEEE Press, 2001.

68. Z. Dang and R. A. Kemmerer. Using the astral model checker for cryptographic protocol analysis. In H. Orman and C. Meadows, editors, *Proc. of the Workshop on Design and Formal Verification of Security Protocols (DIMACS'97)*, 1997.

69. A. Datta, A. Derek, J. Mitchell, and D. Pavlovic. A derivation system for security protocols and its logical formalization. In *Proc. of the 16th IEEE Computer Security Foundations Workshop (CSFW'03)*, pages 109–125. IEEE Press, 2003.

70. R. H. Deng, L. Gong, A. A. Lazar, and W. Wang. Practical protocols for certified electronic mail. *Journal of Network and System Management*, 4(3):279–297, 1996.

71. D. E. Denning and G. M. Sacco. Timestamps in key distribution protocols. *Communications of the ACM*, 24(8):533–536, 1981.

72. T. Dierks and C. Allen. *The TLS Protocol*. Internet Request for Comment RFC-2246, January 1999.

73. W. Diffie and M. Hellman. New directions in cryptography. *IEEE Transactions on Information Theory*, 22:644–654, 1976.

74. D. L. Dill. *Murphi Description Language and Verifier*, 1996. http://verify.stanford.edu/dill/murphi.html.

75. D. Dolev and A. Yao. On the security of public-key protocols. *IEEE Transactions on Information Theory*, 2(29):198–208, 1983.

76. N. Durgin, P. Lincoln, J. C. Mitchell, and A. Scedrov. Multiset rewriting and the complexity of bounded security protocols. *Journal of Computer Security*, 12(2):247–311, 2004.

77. F. J. T. Fábrega and J. D. Guttman. Authentication tests and the structure of bundles. *Theoretical Computer Science*, 283(2):333–380, 2002.

78. F. J. T. Fábrega, J. C. Herzog, and J. D. Guttman. Strand spaces. Research Report 67, The MITRE Corporation, 1997.

79. F. J. T. Fábrega, J. C. Herzog, and J. D. Guttman. Strand spaces: Proving security protocols correct. *Journal of Computer Security*, 7:191–220, 1999.

80. A. Gargantini and E. Riccobene. Encoding abstract state machines in PVS. In Y. Gurevich, P. Kutter, M. Odersky, and L. Thiele, editors, *Proc. of the Abstract State Machines 2000 Workshop*, LNCS 1912, pages 303–322. Springer-Verlag, 2000.

81. D. Gollmann. What do we mean by entity authentication? In *Proc. of the 15th IEEE Symposium on Security and Privacy (SSP'96)*, pages 46–54. IEEE Press, 1996.

82. L. Gong, R. M. Needham, and R. Yahalom. Reasoning about belief in cryptographic protocols. In *Proc. of the 9th IEEE Symposium on Security and Privacy (SSP'90)*, pages 234–248. IEEE Press, 1990.

83. A. D. Gordon and A. Jeffrey. Authenticity by typing for security protocols. *Journal of Computer Security*, 11(4):451–520, 2003.

84. A. D. Gordon and A. Jeffrey. Typing correspondence assertions for communication protocols. *Theoretical Computer Science*, 300(1-3):379–409, 2003.

85. M. J. C. Gordon and T. F. Melham. *Introduction to HOL.* Cambridge University Press, 1993.

86. Y. Gurevich. Evolving algebras 1993: Lipari guide. In E. Börger, editor, *Specification and Validation Methods*, pages 9–36. Oxford University Press, 1995.

87. S. Gürgens and C. Rudolph. Security analysis of (un-) fair non-repudiation protocols. In A. Abdallah, P. Y. A. Ryan, and S. Schneider, editors, *Proc. of the 1st International Conference of Formal Aspects of Security (FASec'02)*, LNCS 2629, pages 97–114. Springer-Verlag, 2003.

88. J. D. Guttman, J. C. Herzog, J. D. Ramsdell, and B. T. Sniffen. Programming cryptographic protocols. In R. De Nicola and D. Sangiorgi, editors, *Proc. of the International Workshop on Trustworthy Global Computing*, LNCS 3705, pages 116–145. Springer-Verlag, 2005.

89. J. D. Guttman, F. J. Thayer, J. A. Carlson, J. C. Herzog, J. D. Ramsdell, and B. T. Sniffen. Trust management in strand spaces: A rely-guarantee method. In D. Schmidt, editor, *Proc. of the 13th European Symposium on Programming (ESOP'04)*, LNCS 2986, pages 325–339. Springer-Verlag, 2004.

90. J. Hastad, R. Impagliazzo, L. Levin, and M. Luby. A pseudorandom generator from any one-way function. *SIAM Journal on Computing*, 28(4):1364–1396, 1999.

91. J. C. Herzog. The Diffie-Hellman key-agreement scheme in the strand-space model. In *Proc. of the 16th IEEE Computer Security Foundations Workshop (CSFW'03)*, pages 234–247. IEEE Press, 2003.

92. C. A. R. Hoare. *Communicating Sequential Processes.* Prentice-Hall, 1985.

93. ISO-7498-2. *Information Processing Systems — Open Systems Interconnection — Basic Reference Model – Part 2: Security Architecture.* International Organization for Standardization, 1989.

94. ISO-8825. *Information Processing Systems — Open Systems Interconnection — Specification of Basic Encoding Rules for Abstract Syntax Notation One (ASN.1).* International Organization for Standardization, 1987.

95. N. Itoi and P. Honeyman. Smartcard integration with Kerberos V5. In *Proc. of the USENIX Workshop on Smartcard Technology*, 1999.

96. R. Jerdonek, P. Honeyman, K. Coffman, J. Rees, and K. Wheeler. Implementation of a provably secure, smartcard-based key distribution protocol. In J.-J. Quisquater and B. Schneier, editors, *Proc. of the 3rd Smartcard Research and Advanced Application Conference (CARDIS'98)*, pages 229–235, 1998.

97. D. Kahn. *The Codebreakers.* Macmillan, 1967.

98. S. Katzenbeisser and F. A. P. Petitcolas, editors. *Information Hiding Techniques for Steganography and Digital Watermarking.* Artech House, 2000.

99. R. Kemmerer, C. A. Meadows, and J. Millen. Three systems for cryptographic protocol analyses. *Journal of Cryptology*, 7(2):79–130, 1994.

100. R. Khamsi. Mathematical proofs getting harder to verify. NewScientist, 19 February 2006. http://www.newscientist.com/article.ns?id=dn8743&feedId=online-news_rss20.

101. J. Kohl and B. Neuman. *The Kerberos Network Authentication Service (Version 5).* Internet Request for Comment RFC-1510, September 1993.

102. J. Kohl, B. Neuman, and T. Ts'o. The evolution of the Kerberos authentication system. In *Distributed Open System*, pages 78–94. IEEE Press, 1994.

103. O. Kömmerling and M. G. Kuhn. Design principles for tamper-resistant smartcard processors. In *Proc. of the USENIX Workshop on Smartcard Technology*, pages 9–20, 1999.

104. T. Leighton and S. Micali. Secret-key agreement without public-key cryptography. In D. R. Stinson, editor, *Proc. of Advances in Cryptography (CRYPTO'93)*, LNCS 773, pages 456–479. Springer-Verlag, 1993.

105. H. Lipmaa. Idea: A cipher for multimedia architectures? In S. Tavares and H. Meijer, editors, *Proc. of the 5th Workshop on Selected Areas in Cryptography (SAC '98)*, LNCS 1556, pages 248–263. Springer-Verlag, 1998.

106. G. Lowe. An attack on the Needham-Schroeder public-key authentication protocol. *Information Processing Letters*, 56(3):131–133, 1995.

107. G. Lowe. Breaking and fixing the Needham-Schroeder public-key protocol using CSP and FDR. In T. Margaria and B. Steffen, editors, *Proc of the 2nd International Conference on Tools and Algorithms for the Construction and Analysis of Systems (TACAS'96)*, LNCS 1055, pages 147–166. Springer-Verlag, 1996.

108. G. Lowe. Some new attacks upon security protocols. In *Proc. of the 9th IEEE Computer Security Foundations Workshop (CSFW'96)*. IEEE Press, 1996.

109. G. Lowe. A hierarchy of authentication specifications. In *Proc. of the 10th IEEE Computer Security Foundations Workshop (CSFW'97)*, pages 31–43. IEEE Press, 1997.

110. G. Lowe. Casper: A compiler for the analysis of security protocols. *Journal of Computer Security*, 6(1-2):53–84, 1998.

111. G. Lowe. Towards a completeness result for model checking of security protocols. In *Proc. of the 11th IEEE Computer Security Foundations Workshop (CSFW'98)*, pages 96–105. IEEE Press, 1998.

112. G. Lowe and A. W. Roscoe. Using CSP to detect errors in the TMN protocol. *IEEE Transactions on Software Engineering*, 3(10):659–669, 1997.

113. D. P. Maher. Fault induction attacks, tamper resistance, and hostile reverse engineering in perspective. In R. Hirschfeld, editor, *Proc. of Financial Cryptography '97*, LNCS 1318, pages 109–121. Springer-Verlag, 1997.

114. A. Mahimkar and V. Shmatikov. Game-based analysis of denial-of-service prevention protocols. In *Proc. of the 18th IEEE Computer Security Foundations Workshop (CSFW'05)*, pages 287–301. IEEE Press, 2005.

115. W. Mao and C. Boyd. Towards formal analysis of security protocols. In *Proc. of the 6th IEEE Computer Security Foundations Workshop (CSFW'93)*, pages 147–158. IEEE Press, 1993.

116. K. McMillan. *Symbolic Model Checking*. Kluwer Academic Publisher, 1993.

117. C. A. Meadows. Formal verification of cryptographic protocols: A survey. In *Advances in Cryptology (Asiacrypt 94)*, LNCS 917, pages 133–150. Springer-Verlag, 1995.

118. C. A. Meadows. The NRL protocol analyzer: An overview. *Journal of Logic Programming*, 26(2):113–131, 1996.

119. A. J. Menezes, P. C. van Oorschot, and S. A. Vanstone. *Handbook of Applied Cryptography*. CRC Press, 1997.

120. J. Millen and V. Shmatikov. Constraint solving for bounded-process cryptographic protocol analysis. In *Proc. of the 8th ACM Conference on Computer and Communication Security (CCS'01)*, pages 166–175. ACM Press, 2001.

121. S. P. Miller, J. I. Neuman, J. I. Schiller, and J. H. Saltzer. Kerberos authentication and authorisation system. Technical Plan Sec. E.2.1, MIT — Project Athena, 1989.

122. R. Milner. *Communicating and Mobile Systems: the Π-Calculus*. Cambridge University Press, 1999.

123. J. C. Mitchell, M. Mitchell, and U. Stern. Automated analysis of cryptographic protocols using Murphi. In *Proc. of the 16th IEEE Symposium on Security and Privacy (SSP'97)*. IEEE Press, 1997.

124. B. Monahan. Introducing ASPECT — a tool for checking protocol security. Research Report 246, HP Laboratories Bristol, 2002. http://www.hpl.hp.co.uk/techreports/2002/HPL-2002-246.pdf.

125. National Bureau of Standards. *Data Encryption Standard*, January 1977. Federal Information Processing Standards Publications, FIPS Pub. 46.

126. R. M. Needham and M. D. Schroeder. Using encryption for authentication in large networks of computers. *Communications of the ACM*, 21(12):993–999, 1978.

127. A. Nenadiċ, N. Zhang, and Q. Shi. RSA-based verifiable and recoverable encryption of signatures and its application in certified e-mail delivery. *Journal of Computer Security*, 13(5):757–777, 2005.

128. T. Nipkow, L. C. Paulson, and M. Wenzel. *Isabelle/HOL: A Proof Assistant for Higher-Order Logic*. Springer, 2002. LNCS Tutorial 2283.

129. H. Orman. *The OAKLEY Key Determination Protocol*. Internet Request for Comment RFC-2412, November 1998.

130. L. C. Paulson. *Isabelle: A Generic Theorem Prover*. LNCS 828. Springer-Verlag, 1994.

131. L. C. Paulson. Mechanized proofs for a recursive authentication protocol. In *Proc. of the 10th IEEE Computer Security Foundations Workshop (CSFW'97)*, pages 84–95. IEEE Press, 1997.

132. L. C. Paulson. Proving properties of security protocols by induction. In *Proc. of the 10th IEEE Computer Security Foundations Workshop (CSFW'97)*, pages 70–83. IEEE Press, 1997.

133. L. C. Paulson. The inductive approach to verifying cryptographic protocols. *Journal of Computer Security*, 6:85–128, 1998.

134. L. C. Paulson. Inductive analysis of the internet protocol TLS. *ACM Transactions on Computer and System Security*, 2(3):332–351, 1999.

135. L. C. Paulson. Relations between secrets: Two formal analyses of the Yahalom protocol. *Journal of Computer Security*, 9(3):197–216, 2001.

136. A. Perrig. Efficient collaborative key management protocols for secure autonomous group communication. In M. Blum and C. H. Lee, editors, *Proc. of the International Workshop on Cryptographic Techniques & E-Commerce (CrypTEC'99)*, pages 192–202. City University of Hong Kong, 1999.

137. R. Rivest. Chaffing and winnowing: Confidentiality without encryption. *CryptoBytes Technical Newsletter, RSA Laboratories*, 4(1):12–17, 1998. http://theory.lcs.mit.edu/~rivest/chaffing.txt.

138. R. Rivest, A. Shamir, and L. Adleman. A method for obtaining digital signatures and public-key cryptosystems. *Communications of the ACM*, 21(2):120–126, 1976.

139. A. W. Roscoe. Model-checking CSP. In A. W. Roscoe, editor, *A Classical Mind, Essays in Honour of C. A. R. Hoare*, pages 353–378. Prentice-Hall, 1994.

140. D. Rosenzweig, D. Runje, and N. Slani. Privacy, abstract encryption and protocols: an ASM model — part I. In E. Börger, A. Gargantini, and E. Riccobene, editors, *Proc. of the Abstract State Machines 2003 Workshop*, LNCS 2589, pages 372–390. Springer-Verlag, 2003.

141. P. Y. A. Ryan. Modelling and analysis of security protocols. Research proposal, Defence Research Agency, 1994.

142. P. Y. A. Ryan, S. Schneider, M. Goldsmith, G. Lowe, and A. W. Roscoe. *Modelling and Analysis of Security Protocols*. Addison-Wesley, 2001.

143. P. Y. A. Ryan and S. A. Schneider. An attack on a recursive authentication protocol: A cautionary tale. In *Information Processing Letters 65*. Elsevier Science, 1998.

144. S. Schneider. Verifying authentication protocols with CSP. In *Proc. of the 10th IEEE Computer Security Foundations Workshop (CSFW'97)*, pages 3–17. IEEE Press, 1997.

145. S. Schneider. Formal analysis of a non-repudiation protocol. In *Proc. of the 11th IEEE Computer Security Foundations Workshop (CSFW'98)*, pages 54–65. IEEE Press, 1998.

146. B. Schneier. *Applied Cryptography: Protocols, Algorithms, and Source Code in C.* John Wiley & Sons, 1994.

147. V. Shmatikov and J. C. Mitchell. Finite-state analysis of two contract signing protocols. *Theoretical Computer Science*, 283(2):419–450, 2002.

148. V. Shoup and A. Rubin. Session key distribution using smartcards. In U. Maurer, editor, *Advances in Cryptology (Eurocrypt'96)*, LNCS 1070, pages 321–331. Springer-Verlag, 1996.

149. P. Smith. LUC public-key encryption. *Dr. Dobb's Journal*, 18(1):44–49, 90–92, January 1993.

150. D. R. Stinson. *Cryptography Theory and Practice.* CRC Press, 1995.

151. P. Syverson. A taxonomy of replay attacks. In *Proc. of the 7th IEEE Computer Security Foundations Workshop (CSFW'94)*, pages 187–191. IEEE Press, 1994.

152. P. F. Syverson. Limitations on design principles for public key protocols. In *Proc. of the 15th IEEE Symposium on Security and Privacy (SSP'96)*, pages 62–72. IEEE Press, 1996.

153. M. Tatebayashi, N. Matsuzaki, and D. B. J. Neuman. Key distribution protocol for digital mobile communication systems. In G. Brassard, editor, *Proc. of Advances in Cryptography (CRYPTO'89)*, LNCS 435, pages 324–334. Springer-Verlag, 1990.

154. URL. AsmL: the Abstract State Machine Language. http://research.microsoft.com/fse/asml/.

155. URL. Cygwin: a Linux-like environment for Windows. http://www.cygwin.com.

156. URL. Isabelle development snapshot. http://isabelle.in.tum.de/devel/.

157. URL. Isabelle download page. http://www.cl.cam.ac.uk/Research/HVG/Isabelle/download.html.

158. URL. Old Isabelle releases. http://www.cl.cam.ac.uk/Research/HVG/Isabelle/download_past.html.

159. URL. Poly/ML: a full implementation of Standard ML. http://www.polyml.org.

160. URL. Proof General: a generic interface for proof assistants. http://proofgeneral.inf.ed.ac.uk.

161. URL. Standard ML of New Jersey. http://www.smlnj.org.

162. VISA International. *Global Smart Card Growth Continues as Visa Surpasses 40 Million Mark*, 2001. http://usa.visa.com/about_visa/newsroom/press_releases/nr98.html.

163. VISA International. *3-D Secure Introduction*, 2006. http://partnernetwork.visa.com/pf/3dsec/.

164. M. Wenzel. Isar — a generic interpretative approach to readable formal proof documents. In Y. Bertot, G. Dowek, A. Hirschowitz, C. Paulin, and L. Thery, editors, *Proc. of the 12th International Conference on Theorem Proving in Higher Order Logics (TPHOLs'99)*, LNCS 1690, pages 167–184. Springer-Verlag, 2001.

165. K. Winter. Towards a methodology for model checking ASM: Lessons learned from the FLASH case study. In Y. Gurevich, P. Kutter, M. Odersky, and L. Thiele, editors, *Proc. of the Abstract State Machines 2000 Workshop*, LNCS 1912, pages 341–360. Springer-Verlag, 2000.

166. T. Y. C. Woo and S. S. Lam. Authentication for distributed systems. *Computer*, 25(1):39–52, 1992.

167. T. Y. C. Woo and S. S. Lam. A semantic model for authentication protocols. In *Proc. of the 12th IEEE Symposium on Security and Privacy (SSP'93)*, pages 178–194. IEEE Press, 1993.

168. G. Zhou and D. Gollmann. Towards verification of non-repudiation protocols. In J. Grundy, M. Schwenke, and T. Vickers, editors, *Proc. of the International Refinement Workshop and Formal Methods Pacific*, pages 370–380. Springer-Verlag, 1998.

169. J. Zhou. *Non-repudiation in Electronic Commerce*. Artech House, 2001.

170. J. Zhou, R. H. Deng, and F. Bao. Evolution of fair non-repudiation with TTP. In *Proc. of the 4th Australasian Conference on Information Security and Privacy*, LNCS 1587, pages 258–269. Springer-Verlag, 1998.

171. J. Zhou and D. Gollmann. A fair non-repudiation protocol. In *Proc. of the 15th IEEE Symposium on Security and Privacy (SSP'96)*, pages 55–61. IEEE Press, 1996.